D1806147

# REFRACTORIES AND THEIR USES

KENNETH SHAW

# Refractories and Their Uses

APPLIED SCIENCE PUBLISHERS LTD

LONDON

APPLIED SCIENCE PUBLISHERS LTD
RIPPLE ROAD, BARKING, ESSEX, ENGLAND

ISBN 0 85334 541 4

With 20 Tables and 68 Illustrations

© 1972 KENNETH SHAW

All rights reserved. No part of this publication may be reproduced,
stored in a retrieval system, or transmitted in any form or by
any means, electronic, mechanical, photocopying, recording, or
otherwise, without the prior written permission of the publishers,
Applied Science Publishers Ltd, Ripple Road, Barking, Essex,
England

Printed in Great Britain by Galliard Limited, Great Yarmouth, Norfolk, England.

FOR

MARY AND CATHERINE

# Preface

Workers in many industries need to know about heat-resistant materials. They include design engineers, materials technologists, chemical engineers, technicians, builders and ceramists. In order to practise their own disciplines they must consult the literature outside it. This book is written particularly with these readers in mind, since most of the up-to-date books on refractories are written for the refractories manufacturer or for the specialist who is assumed to have a wide knowledge of the basic principles of refractories technology.

I have attempted to present a concise summary of the existing knowledge of the subject which I hope will appeal to non-specialists and students. The volume may also be useful to those workers inside the refractories industry who seek an overall view of the technology. The references to recent Soviet developments, in which the author has a special interest, may also be of value to those engaged in research into refractory materials and their applications.

The literature dealing with the production and use of refractories is vast. Most industries find it necessary to use heat of some intensity or other and this in turn requires the employment of heat-resistant containers. Engineers have designed hundreds of different types of furnace linings. If we take into account the various ways in which these linings may react with the metals, ores, clays, cements, glass and slags which are processed in furnaces, it becomes apparent why the subject of refractories science is so vast.

This book is divided into three sections. The first summarises the basic principles of the production of refractories and describes their general properties. In the second section the individual types of heat-resistant materials are discussed with special reference to the differences in their physicochemical properties and the reasons for some being preferred to others in certain applications. The third section describes the use of refractories under the headings of various industries where large quantities of heat-resistant bricks, mortars, castables, coatings and cements are consumed. In this third section close attention has been paid to citing useful references from the literature of Europe, the USA and the USSR, because it is in the application of refractories, especially new ones, that the non-specialist reader, who may be unfamiliar with the sources of information, will find the greatest difficulty in obtaining information for answering specific inquiries and problems.

The behaviour of refractory linings in service is a huge field for research. As new processes for applying heat to raw materials are developed and put to use, new materials have to be found to meet the increased demands placed on the furnace linings. Thus, our knowledge of the reactions between linings and furnace charge needs to be continuously reviewed. The reader will find that many of the papers cited at the ends of the chapters in the third section contain information about large operational furnaces and kilns. By following up these references it should be possible to find detailed data about the related theory and laboratory work which preceded the industrial trials and examinations of demolished furnaces, information about which has formed the basis of much of this third section of the present volume.

A large number of technical colleges and universities are now offering courses at various levels which include heat technology, ceramics, or refractories. This book may be useful to students taking many of these courses, especially for those who, while not specialising in ceramics or refractories, need some acquaintance with refractories and their uses.


Brixham                                    KENNETH SHAW
February, 1972

# Contents

ix

## SECTION II: TYPES OF REFRACTORIES

*SECTION I*

# SELECTING REFRACTORIES

*Chapter 1*

# What is a Refractory Material?

A material can be described as 'refractory' if it can stand up to the action of corrosive solids, liquids, or gases at high temperatures. Refractory materials are used to make furnaces, kilns, stoves, driers, and critical parts of jet engines, missiles and aerospace craft.

The various combinations of conditions in which refractories are used make it necessary to manufacture a range of refractory materials with different properties. This involves selecting raw materials with specific characteristics, processing them and finally fabricating them into shapes with the desired combinations of properties to meet the particular demands of a given environment. For example, the relatively clean atmosphere inside an electrically fired tunnel kiln used for the manufacture of bone-china needs quite a different type of refractory lining from that which is demanded in the construction of the hearth of a blast furnace. Firebricks used in a domestic fireplace, burning house coal, are obviously unsuitable for use in the tank of a glass furnace, melting glass at temperatures around 1 500°C.

## CLASSIFICATION OF REFRACTORIES

The classification of refractories (*see* Chapter 5) and the specifications that have been drawn up in attempts to standardise their production and use are therefore an important aspect of any attempt to describe these materials. Broadly speaking we may put any refractory or heat-resistant material into one of three classes, based on the temperature range in which it will fuse:

1. Refractory materials fused at 1 580–1 780°C.
2. Highly refractory materials fused at 1 780–2 000°C.
3. Super-refractory materials fused at above 2 000°C.

Within each group there are many sub-classes which are governed by chemical and mineralogical composition, purpose, and physical properties

3

such as density, porosity, etc. Some of these sub-groups (discussed more fully in Chapters 6–17) are: silica or dinas materials; aluminosilicate such as firebrick and mullite brick; magnesite and other basic bricks, and carbon and silicon carbide.

Most of these sub-classes are 'bulk' or tonnage refractories used in large quantities for furnace and kiln construction. Other sub-classes contain special, high-cost materials which, because of their scarcity and difficulty of fabrication, are used in limited quantities, *e.g.* certain oxide ceramics such as beryllia and uranium-oxide refractories.

## SHAPES AND FORMS

The furnace or kiln builder obtains his refractories in many different forms. The common firebrick is probably the most easily recognisable refractory material. However, refractories are made in a very wide variety of shapes to suit the designer's requirements. Large blocks are made if it is necessary to reduce the number of joints in order to avoid the penetration of molten slags and glass into the brickwork. Plugs, tubes and tundishes made of refractory materials are commonly used in the smelting and casting of metals. In more sophisticated technology such as aerospace, complex shapes are fabricated for critical application in missile and spacecraft design.

Most refractories are shaped and fired into hard, strong, resistant articles that can be safely transported to the places where they are to be used. Other types of refractories are produced in powder or slurry form. For example, open-hearth practice in the iron and steel industry has necessitated the use of vast quantities of metallurgical powder (magnesite and chrome-magnesite) for mending and restoring furnace hearths and other parts: a process known as fettling. Many furnaces are now built, not with bricks and mortar, but with refractory concretes which can be poured into wooden shuttering to make the walls and roofs etc., thus eliminating the need for skilled bricklayers and also saving the time that used to be spent building up the structure by traditional bricklaying methods. For this, castable mixtures of refractory cements and grogs (chamotte) are made, which, when mixed with water and allowed to set, produce monolithic structures with properties similar to those built of fired bricks with mortar.

The density of refractory materials is an important property and to some extent refractories may be classified simply with respect to density. High-density refractories made by applying high pressures to the raw materials that have been ground and processed with clay or other bond, or even by melting the raw materials and casting them into moulds in much the same way as steel or glass is formed, can be virtually pore-free. At the other end of the scale, by using foaming agents, or by adding organic materials (such as sawdust or cork which burn out in firing) to the refractories, the manufacturer can produce lightweight, highly porous materials that are 'full of air' and therefore ideally suitable as heat insulation. Air does not conduct heat very well, thus by trapping air in the clay or other refractory materials it is possible to produce satisfactory insulators. The porosity range of refractory materials available commercially runs from less than 3% to more than 85%.

## CLASSIFICATION BY FABRICATION METHOD

Another way of classifying refractories is by the manner in which they are shaped. Briefly, refractories are made by selecting suitable raw materials, processing them to remove unwanted impurities such as iron oxides and fluxes, crushing, grinding or milling to the most suitable particle sizes, depending on the ultimate purpose, and then putting these particles together again so that the desired structure is built up. The final process in fabricating a refractory material is normally that of heat treatment, which may take the form of firing in a kiln, sintering under pressure, or melting and casting into moulds.

During heat treatment the particles of raw materials are bonded together either by means of self reactions, initiated and carried on by the heat applied in the furnace, or with the aid of a bond in the form of a glassy material specially added for the purpose. Clay acts as a glassy bond in many common refractories since at high temperatures it fuses and sticks together the particles of grog (chamotte) that form the structure. The manner in which this structure forms and the devices used by the refractories technologist constitute some of the most important aspects of the manufacture and use of heat-resistant materials. On these processes depend the life and behaviour of the materials in service, that is, when they are built into a furnace lining and come into contact with hot gases, slags, molten glass, metals, cement clinker or any of the other materials that man subjects to heat treatment for use in his industries.

## HOW ARE THEY SHAPED?

The following fabrication techniques are used in the refractories industry.

1. Cutting up blocks of natural stone-like materials and using these in furnace linings, e.g. diatomite.
2. Plastic moulding to produce a wide variety of shapes, either by hand or by machine.

3. Slip casting, a process which involves grinding raw materials in water or other media to yield cream-like slurries which can be cast into moulds made from plaster of Paris or other absorbent materials. The mould sucks up the water, leaving the cast to be dried and fired.
4. Dry or semi-dry ramming and pressing. The 'dry' materials contain small amounts of moisture and plasticising additives such as clay, or organics which burn out on firing. A wide range of presses is used.
5. Fusion casting, which consists in melting the refractory raw materials and pouring the fusions into moulds.
6. Hot pressing, which consists in heating the raw materials until they start to fuse, that is, become plastic in the hot state, and then applying pressure in order to densify the bodies.

Various other specialised techniques, such as thermoplastic methods involving the use of temporary waxy bonds, such as paraffins, are also used. More details of the principles of ceramic fabrication processes will be found in Chapter 4.


## SELECTING REFRACTORIES

The choice of refractory material for a given application will be determined by the type of furnace or heating unit and the conditions prevailing, *e.g.* the gaseous atmosphere; the presence of slags; the type of metal charge and the degree of purity required in the final product. It is clear, therefore, that temperature is by no means the only criterion and indeed the degree of refractoriness, or the fusing point of a material, will give only a very limited indication of how it will stand up in service in industrial furnaces.

It is customary to characterise refractories by a combination of properties, the testing of which is the subject of continuing research and development (*see* Fig. 1.1) because as yet no perfect test has been devised to provide all the necessary information about the service of a refractory material in a furnace. In other words, 'refractoriness' in practice is a multi-dimensional characteristic. The properties in which a prospective user of refractories should be interested are as follows:

Refractoriness (reported either as the softening temperature, or the pyrometric cone equivalent, PCE); the refractoriness under load, or the temperature at which the material deforms under a certain load; the volume stability, that is, whether it shrinks or expands when heated, and by how much; the thermal conductivity; the slag and molten metal resistance; thermal-shock resistance, also termed spalling resistance, a property which indicates whether the material will crumble or flake when subjected to alternating heating and cooling; the mechanical strength, and the porosity.

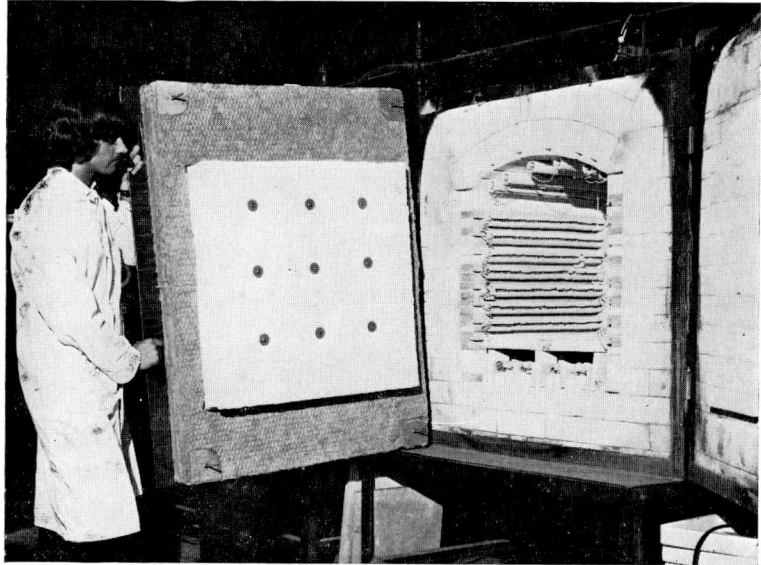

Fig. 1.1. Specially designed furnaces are used to simulate the service conditions of refractories. Here the high-temperature shrinkage of a ceramic-fibre blanket is being measured. Note the densely set electrical-resistance heating elements inside the furnace. *Photo:* Courtesy British Ceramic Research Association.

Other properties which may be important for certain applications include electrical conductivity, gas permeability, external appearance and structure. Each of these properties is discussed in Chapter 2.

## ACID AND BASIC REFRACTORIES

In the processing of metals, cements, glass and other materials it is common for the furnace charge, that is, the material being processed, to react chemically and physically with the refractory lining. Indeed, the degree of this reaction will govern the life of the lining, hence that of the furnace. When the charge, or the slags produced, or merely the furnace gases, have eaten away or reacted with the lining to the extent that holes are formed and the structure is in danger of collapse, the furnace has to be repaired or demolished.

In many furnaces the way in which the charge materials react with the refractories in the lining depends on the chemical nature of the reactants, that is (putting it very simply) whether the lining and charge are chemically acid or basic. Again, very simply, if the lining were acid, as for instance it would be if it were made from silica (dinas) bricks and the furnace charge were basic, as it would be if the slags in it contained a high percentage of lime and magnesite, then the degree of chemical reaction between the two would be intense. This is simply because acids react vigorously with bases. The lining would be operating in ideal conditions for early failure because of high temperatures and fast chemical reaction rates. The charge would very rapidly react with, dissolve, and eat away the lining, so that eventually there would be no lining left at all: merely charge materials. Obviously, the furnace would be stopped long before this stage was reached, and in any case furnace designers take care to choose lining materials that do not react so readily with the charge, although it must be stated that charge-lining reactions are often quite intense, and their study inevitably constitutes an important section of refractories technology. Some of these reactions are discussed in Section III, dealing with the applications of refractories, and in Chapter 2.

Acid refractories, those that react with basic slags and fusions, include silica brick (dinas), fireclay, and some other aluminosilicate refractories.

Basic refractories include magnesite, magnesite-chromite, and various combinations of magnesia and chromite.

It will be apparent that before choosing his refractory lining the furnace designer must first understand the chemical nature of the materials being processed in the furnace. This is so because as yet no refractory has been developed which combines all the properties needed for stable operation under any set of conditions likely to be encountered in industrial usage.

Since each type of refractory possesses its own specific properties the matter of choice is of the utmost importance. It is customary for the user to collaborate with and be advised by the manufacturer of refractories and a feature of the modern refractories industry is the increasing extent of this collaboration. In many countries large users of refractories, such as the iron and steel industry which uses about $70\%$ of the total volume of refractories made in the world, have their own refractories divisions and these are responsible for the development and testing of new types of materials for furnace linings.

In choosing a refractory material, a furnace operator will be guided first by the most critical operating factor and then he will need to compromise in order to balance all the properties of the refractories available. For example, silica bricks have a very high refractoriness under load at relatively high temperatures; they can be heated within $10°C$ or so of their point of failure and are thus suitable for making the roofs of open-hearth steel furnaces and glass-tank furnaces. However, silica bricks suffer from an unsatisfactory thermal-shock (spalling) resistance at certain temperatures, especially when they are being warmed up from cold, and they also have a low slag resistance, especially in the presence of highly basic slags, as explained above. In using silica for roof structures it is therefore necessary to take great care in warming up to avoid cracking and flaking and then to make sure that the maximum temperature at which the roof is working will not be accidentally exceeded.

## STANDARD SPECIFICATIONS

Every major industrialised country in the world has issued standard specifications for refractory materials. These standards describe testing methods and classifications. It is some indication of the long road yet to be travelled in refractories technology that there exist many marked discrepancies and differences of opinion about the methods of testing refractories and even the purpose of the tests.

*Chapter 2*

# Properties of Refractories

The properties of refractory materials depend on their chemical and mineral composition, and on their structure. By structure is meant the number and sizes of the grains and pores, the manner in which the pores are arranged in the bulk of the material, and the physical strength of the body. In turn, structure is closely allied to the method used to shape the raw materials into products, and, indeed, fabrication processes are probably the key to any future improvement in the service life of refractories, for man appears to be reaching the ultimate level in the economic purification of most of the raw materials available for furnace construction.

Refractories possessing the same substance composition but fabricated in different ways to yield different structures may possess quite different physical properties, *e.g.* wet moulded firebrick may be highly porous, whereas a dry-pressed brick, pressed at very high pressures and having the same composition, can be very dense.

As stated in Chapter 1, refractories are characterised by a range of properties, all of which may be subjected to some kind of testing. Each of these properties will now be considered.

## REFRACTORINESS

It is common practice to report the 'refractoriness' of a material in terms of temperature, that is, the temperature at which the material fuses, softens, or melts. However, in contrast to pure compounds and metals, which have distinct, easily definable melting points, refractories soften or fuse over a range of temperature. It is therefore misleading, and normally of limited usefulness, to assign a melting point to refractories. The temperature at which a refractory such as firebrick or silica brick (dinas) fuses depends on the test or service conditions. For standardised conditions, therefore, we may say that the refractoriness of a material is its power to resist, without fusing, the effect of high temperatures. However, simple refractoriness tests alone cannot be used for specifying the temperatures at which the materials can be safely used.

10

In practice refractoriness is measured by comparing the behaviour of a small cone of the material with that of other, standard cones which have known softening points when they are heated under carefully controlled conditions. The result is reported as the pyrometric cone equivalent (PCE) of the refractory, which means simply the number of the cone that most nearly corresponds in appearance to the test cone at the time of softening. Figure 2.1 shows a set of test cones, from which the principle of the method can be understood. The rate of heating, the furnace atmosphere, the cone sizes and the methods of mounting are all specified. Each cone is numbered, and tables exist showing the temperatures that are equivalent (approximately) to cone numbers.

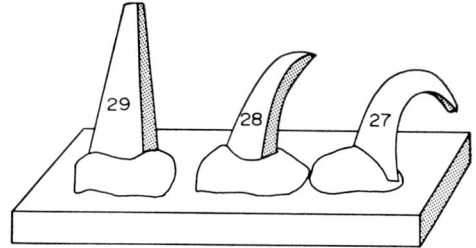

FIG. 2.1   Pyrometric cones during heating (see text).

When a refractory material is heated a liquid forms in it, and as the fluidity of this liquid increases the material starts to slump and deform; it loses its cohesiveness. The nature of the liquid formed and its ultimate effect on the refractory depends on the chemical and mineralogical compositions, the particle size, the heating rate, the shape and size of the specimen (or brick), and the atmosphere surrounding the refractory, that is, oxidising, reducing, or neutral. This, simply, is why the term 'refractoriness' is not equivalent to a physical constant, and it only has meaning if the test conditions are maintained constant.

The heating rate is important because if it varies the PCE (refractoriness) will be reported as being different from what it would be for a different heating rate. A slower heating rate causes the PCE to be undervalued. Thus, the test to some extent is concerned with the amount of heat-work (the product of temperature and time) done on the refractory (Fig. 2.2).

The sizes of the particles of the various materials in the refractory (chamotte, silica, clay substance, etc.) are also important, since usually the finer they are the faster the rate of softening, other conditions being equal. However, this is also governed by the overall structure (porosity, density and the relationship of the pores and their sizes) of the material, which is discussed below.

Fig. 2.2.   Special refractories are needed in the construction of laboratory furnaces which are used for testing ceramics and refractories. This shows a small laboratory unit for temperatures up to 1 200°C. *Photo:* Courtesy Gallenkamp.

Furnace atmosphere is important because impurities such as iron oxide behave quite differently in oxidising conditions from the way they behave in reducing conditions. For heavily iron-contaminated firebrick, for instance, tests made in reducing conditions will show much lower PCE results than if the tests were done in an oxidising atmosphere, since ferrous oxide, FeO (the lower, or reduced, form of iron oxide), more readily produces low-fusing compounds with the other ingredients of the material than does ferric oxide $Fe_2O_3$. Refractoriness tests are usually done in oxidising atmospheres unless special data are required.

## HOT STRENGTH UNDER LOAD

The resistance of a material when it is heated to high temperatures under load is of more practical significance than its refractoriness, if the latter is reported merely as its pyrometric cone equivalent or fusing point. In service, that is, when it is built into a furnace or other heating unit, the material may face the action of various types of mechanical load: tension, compressions, scaling, abrasion, and also thermal shock. Very high pressures may be exerted by a furnace wall on the bottom rows of the brick. For a firebrick structure, the pressure has been calculated to be 0·18 kg/cm² for every metre of wall height. Depending on the type of furnace and the materials, therefore, the pressures will range from 1·0 to 10 kg/cm².

The refractoriness-under-load is usually tested under a load of 2 kg/cm² for dense refractories such as firebrick and 1 kg/cm² for insulating (porous) materials, by heating a test block at a gradually rising temperature in a silicon carbide resistance or other suitable furnace. The temperatures at which the specimens start to deform or sag, and eventually fail (usually due to shearing), are then reported. In British and American practice it is common procedure to draw a graph to show the rate of shearing or sagging of the material with temperature rise.

A refractory specimen undergoing testing may consist of a mixture of crystals and glass, and the mixture can be said to be elastoplastic, that is, it must be subject to a certain amount of stress before flow commences. Once this flow does commence, however, the material behaves much the same as a glass, and it deforms like a glass. The deformation of refractories at high temperatures is mainly governed by the chemical and phase compositions, that is, the same factors as determine the refractoriness. Other important factors are the viscosity of the melt (glass phase) and the porosity of the material, though the latter is of much less importance than the chemical and phase compositions. For a given composition, the less dense the refractory the lower its refractoriness-under-load value. Once deformation commences, the porosity has little effect.

### Failure under load

The precise manner in which a material fails under load at high temperatures varies with the type of material, e.g. silica brick (dinas), firebrick (including high-alumina products), and magnesite (basic). Many firebricks show a gradual slumping kind of deformation, while silica brick seems to be perfectly stable one second and then suddenly collapses the next. The explanation lies in the fact that in silica brick the heat-resistant constituent takes the form of a strong crystal entanglement that is relatively insoluble in the glassy phase formed from the traces of alumina and iron in the silica brick, whereas in the firebrick the main mass is granular, with about half of it comprising a glassy phase which fuses over a wide temperature range.

Many refractories fail under load at temperatures much lower than the fusing point (PCE) of the material when it is tested without loading. Silica brick, however, has a PCE of 1 710–1 720°C and a refractoriness-under-load (2 kg/cm²) of 1 650–1 660°C—a difference of only 50–60°C. It is therefore an excellent material for suspended roof structures operating at high temperatures, where the temperature cycle can be strictly controlled. Silica brick contains up to 15% vitreous material, the rest being crystalline tridymite, quartz, and cristobalite (forms of silica). The tridymite forms an intergrowth of crystals which neutralise the potential damaging action of the 15% fusible constituent (glass). It is only when the tridymite starts to fuse, rather suddenly at 1 650–1 670°C, that the silica brick fails. Failure occurs over a range of 10–15°C. Another reason for this particular behaviour in refractories such as silica brick is the high viscosity of the liquid phase.

Turning to a typical basic refractory—magnesite brick—it is noted that the magnesia crystals cannot develop intergrowths but are stuck together by a monticellite (magnesium–calcium–silicate) bond which starts to soften at 1 450°C and fuses over the range 1 450–1 550°C. The magnesia crystals in the brick are only slightly soluble in the monticellite fusion; therefore, as the temperature rises the refractoriness does not rise. The viscosity of the liquid falls rapidly with the rise in temperature and sudden failure occurs in magnesite bricks. This mechanism also explains the big difference (700°C) between the PCE of magnesite and its refractoriness-under-load. Manufacturers therefore aim at reducing the amount of glassy material and increasing its viscosity at service temperatures.

When firebrick is heated under load its content of liquid increases, but so too does the viscosity of this liquid owing to the solution in it of silica and alumina from the firebrick grog (chamotte). This explains the smooth, plastic type of deformation under load of firebrick and other aluminosilicate refractories. Firebrick containing about 5% flux impurities ($CaO$, $Fe_2O_3$, alkalis, etc.) softens at 1 350–1 400°C, and fuses at 1 570–1 600°C.

From all this it is apparent that the type of high temperature bond existing in refractories determines the nature of failure under load. The two main types of bond are:

(a) direct bond between intergrowing crystals (*e.g.* silica);
(b) ceramic bond of a fusible glassy phase existing between crystals (firebrick and magnesite).

## CREEP

The term 'creep' is used in refractories technology to describe the heat-activated process of the plastic deformation of crystals under stress. It occurs very slowly in a manner that is typical of the diffusion processes in which the decisive effect is exerted by thermal activation. Creep is generally considered as plastic deformation due to shear, resulting from an applied load, and diffusion, the role of which is enhanced with temperature. Shear in crystalline materials occurs by dislocation movement.

Many theories have been proposed for the precise mechanism of creep. They are based chiefly on the theory of plastic flow and small elasto-plastic deformations. The subject is complex and is still being extensively researched. A knowledge of the creep behaviour of all forms of refractories is obviously important in evaluating the constructional properties of refractories designed for service under load at high temperatures, since it is necessary to know the maximum permitted loading at the appropriate temperature and the extent of the deformation, if any.

In view of the differences of opinion about the precise nature of creep in refractories, several methods have been proposed for studying it. Most contemporary researchers are concentrating on a method in which the creep is determined under the axial loading of specimens, while making continuous measurements of their dimensions, under constant temperature. The load is varied.

The importance of creep in refractories is such that many researchers have stated that creep more properly reflects the failure of refractories in furnaces than the refractoriness-under-load at a fixed load of 2 $kg/cm^2$, determined by standard methods, since the creep test approaches more closely the service conditions in actual furnaces.

Cases are common of refractories withstanding loads of 2 $kg/cm^2$ at very high temperatures for a short period, but proving to be quite unsuitable for prolonged service at much lower temperatures. Information on the creep of refractories is therefore of great importance in designing furnaces. In choosing refractories for particular applications, the furnace designer must know their creep values at various temperatures, specified by a single method, since only then can he evaluate the refractories' quality from this criterion.

## VOLUME STABILITY AT HIGH TEMPERATURES

### Permanent changes

Refractory materials tend to shrink or expand when heated in service, not only because of the normal physical behaviour of any constructional material, but also owing to mineral inversions, physicochemical reactions, sintering, and other effects that occur during firing. In service, refractories usually have to withstand temperatures higher than those used to fire them in the manufacturer's kiln.

The 'after-contraction' and 'after-expansion', that is, the permanent volume changes occurring during heat soaking, are determined by heating the materials in a test furnace at various temperatures (say 1 350 or 1 410°C) with a specified heating cycle, and measuring the dimensions before and after the test.

Any volume change tends to impair the strength of refractories in service, although sometimes ladle and other firebricks are deliberately made to expand (bloat) in service in order to produce a tighter lining that will better withstand slag and metal attack at the mortar joints.

Suspended refractories such as those built into roofs must not shrink if the roof structure is to remain intact.

Expanding refractories are usually unsuitable for most applications since the expansion sets up stresses, causing bowing and possibly bursting of the refractory linings.

The ideal, therefore, would be a perfectly volume-stable material, but in reality we find that firebrick, chrome–magnesite, forsterite, and magnesite, as well as most aluminosilicate and aluminous refractories, do shrink in service. Silica brick, on the other hand, expands when heated owing to the critical silica transformations (*see* Section II, Chapter 8). Highly siliceous materials, or those containing specially added expanding aggregates, such as some types of kyanite, may be volume-stable, or show a slight expansion. In these mixtures the shrinkage of the fireclay is compensated for by the expansion of the kyanite.

**Reversible thermal expansion**
A distinction should be made between the permanent after-contraction of a refractory and the reversible thermal expansion. The reversible thermal expansion of refractories, $\alpha$, and the subsequent contraction on cooling are important properties, not only to the kiln designer and builder who must consider them when designing joints, but also to the refractory technologist and user who can predict certain other properties from $\alpha$-values, *e.g.* spalling resistance (q.v.). The $\alpha$-value determines the level of the stresses formed in a lining, especially when sudden temperature changes occur. Expansion factors depend on the composition of the material and the temperature. The structure, density and strength do not greatly affect the linear expansion coefficient. The expansion curve may alter if the firing temperature of the refractories is altered.

The numerical value of the mean coefficient for the majority of refractories is low. Some examples are as follows (for the temperature range 20–1 000°C):

| | |
|---|---|
| Firebrick | $4.5–6.0 \times 10^{-6}$ |
| Alumina (99% $Al_2O_3$) | $8.0–8.5 \times 10^{-6}$ |
| Silica brick | $11.5–13.0 \times 10^{-6}$ |
| Magnesite | $14.0–15.0 \times 10^{-6}$ |
| Chrome–magnesite | $10.0 \times 10^{-6}$ |

To convert coefficient values to practical dimensions it is necessary to multiply the α-value by the temperature at which the refractory is to be used, and the product by 100. For example, the linear expansion of alumina at 800°C is $8\cdot0 \times 10^{-6} \times 100 = 0\cdot64\%$. Thus, one metre of alumina lining would be elongated by 6·4 mm when heated to 800°C.

The volume coefficient of expansion is taken to be triple that of the linear coefficient.

Several complicated methods are used to measure reversible expansion, e.g. the interferometer, fused quartz-tube, dilatometers, and comparator methods.

The thermal expansion of silica brick is of special interest because a heating curve plotted against the volume or linear changes in the brick indicates quite clearly the important crystal inversions that are taking place. This aspect of silica refractories technology is discussed in Section II, Chapter 8.

## THERMAL-SHOCK (SPALLING) RESISTANCE

Thermal-shock resistance (or spalling resistance) is the capacity of refractories to retain their original form without cracking, splitting, or flaking when subjected to sudden temperature changes. It is a vital property for most refractories and users must have as full an understanding of it as possible (Fig. 2.3). A refractory may spall in a variety of ways owing to one or a combination of causes, including the following:

(a) differences in thermal expansion factors between layers within the brick, brought about by service conditions, e.g. slagging, and structural changes through gas permeation, leading to catalytic reactions and changes in density within the structure;

(b) temperature gradients in the refractory;

(c) compression in the lining owing to volume changes;

(d) other stress-inducing mechanisms.

Providing a body is isotropic and homogeneous no stresses will arise in it during free thermal expansion. However, refractory linings are not in practice allowed to operate under these ideal conditions, and real structures invariably possess stresses.

Compressive stresses develop because dimensional changes are prevented from occurring during heating, for the material is trying to expand. Tensile stresses develop during cooling for the opposite reason. If these stresses exceed the shear or tensile strength of the refractory, cracks develop.

The structure, and especially the grain-size distribution of the grog (chamotte), is of special importance in spalling studies. Methods have been

FIG. 2.3.   The thermal-shock (spalling) resistance of refractories is difficult to test in the laboratory. Panel tests, as shown here, are more satisfactory than water-heat cycling of small test cubes. *Photo:* Courtesy British Ceramic Research Association.

developed to induce a microcracked structure in fireclay refractories in order to improve spalling resistance. Since theoretical considerations, involving the application of formulae that take into account tensile and shear moduli, elastic properties and their correlations with mineral and chemical compositions, are as yet only of limited practical interest, we shall restrict the present discussion largely to empirical knowledge.

The requirements of a highly spalling-resistant refractory are: (1) a structure that will readily dissipate any thermal stresses; (2) a material that will retain its structure in service.

The size and shape of the article will also affect the spalling resistance, and the results obtained from testing will therefore vary if standard samples are not used.

No fully satisfactory theory of thermal-shock resistance has yet been evolved, but in general if dimensional and structural features are ignored as well as test conditions the property can be specified by the thermal resistance factor $K$, calculated from the formula

$$K = \frac{\lambda\sigma}{c\gamma\alpha E}$$

where $\lambda$ is the thermal conductivity coefficient; $\sigma$ is the breaking strength;

$c$ is the specific heat; $\gamma$ is the bulk density; $\alpha$ is the thermal expansion coefficient; and $E$ is Young's modulus.

This theoretical formula, although of little practical use (even for comparisons, since it ignores so many real conditions), does indicate some useful relationships. For instance, thermal-shock resistance rises with an increase in thermal conductivity and mechanical strength, but falls with a rise in the thermal expansion coefficient and Young's modulus. The specific heat and bulk density are constants for a given refractory.

Numerous tests have been devised to check the spalling resistance of refractories, the simplest being to heat specimens or whole bricks in a laboratory furnace to, say, 1 000–1 300°C, and remove them in the hot state and drop them into cold water, or simply on the floor of the laboratory.

The number of times they will stand this treatment without spalling is a rough measure of their thermal-shock resistance. Various heating and cooling rates are recommended, depending on the type of material.

Another method, more closely imitating real conditions in furnaces, involves heating test blocks from one end with a hotplate and measuring the rate of heating needed to crack the piece. This test is also dependent on the test piece size and shape.

Panel tests in which bricks are assembled in the form of one side of a test furnace and heated as they would be in a real furnace are very satisfactory methods of testing for thermal shock (*see* Fig. 2.3). Unfortunately they are laborious, costly, and take a long time.

As mentioned above, structure has an important effect on the spalling resistance of refractories, particularly aluminosilicate refractories. One method of testing spalling resistance is to measure the destructive temperature gradients set up in the material. The level of the gradient diminishes when the porosity of the material is reduced. Nemets and colleagues* studied the thermal-shock resistance of firebrick and developed a rule for the change in spalling resistance during re-orientation of microcracks developed in the structure. The level of the spalling resistance of firebrick obtained by a destructive temperature gradient was in agreement with industrial tests of steel plant refractories with a high concentration of microcracks orientated around the boundaries of the coarse grains of grog. These authors state that any evaluation of the spalling resistance of heterogeneous refractories by the known calculation criterion should take into account the features of the structure.

Studying the spalling resistance of forsterite and chrome–forsterite refractories, other Soviet authors† found that the property is governed by the macro- and microstructure, including the distribution of the large and

* I. I. Nemets *et al.*, *Ogneupory*, No. 4, pp. 56–59, April (1969).

† V. A. Bron *et al.*, *Ogneupory*, No. 9, pp. 50–57, September (1969).

fine pores. In forsterite refractories the microcracks develop into larger cracks which spread over a large area, which is a cause of the reduced spalling resistance. In chrome–forsterite refractories the development of microcracks is localised in small areas which retards the breakdown process. It is stated that the residual strength may be used to satisfactorily characterise the thermal-shock resistance of forsterite refractories, determined in the prescribed manner. Forsterite refractories are spalling resistant during cyclic heating up to 800°C, and chrome–forsterite up to 1 300°C. After being subjected to multilateral cyclic heating the products have a lower residual strength than after single-sided heating.

## SLAG AND METAL RESISTANCE

In the metal industries many refractory linings have to withstand the splashing and washing action of molten metals and slags, not to mention furnace gases and fuel ash, and the rubbing action of raw materials. At the high temperatures prevailing these corrosive agents react with the refractory materials of the lining, and the study and understanding of these reactions constitutes much of refractories science.

Slag attack is particularly important. The structural strength of the refractory may be critically reduced by the solvent action of liquid slags. Two processes may occur: corrosion, which is a chemical reaction; and erosion, which is the process of breaking and washing away the refractory materials by molten slag.

These two important processes of refractory destruction depend on many factors, including: the temperatures and temperature gradient (the slag will be hotter nearer to the furnace core than inside the structure of the lining); the chemical and mineral composition of the slags and linings; the structure (porosity) of the lining and the sizes and forms of its pores; whether or not the lining is wetted by the slag; the viscosity of the slag; the composition of the gaseous atmosphere in the furnace and the rate of reaction between slag and lining, and the speed at which slag-lining reaction products are removed.

Sometimes slag compositions are designed to foster slag attack on the lining with the aim of producing a layer of resistant material that protects the lining from further attack (*e.g.* in rotary cement kilns, q.v.). However, in general the aim is to use refractory linings that are not wetted by, and will not react with, slag. It is for this reason that acid refractories (*e.g.* silica brick) are not used in conjunction with highly basic slags, for the reaction product would consist of a low melting compound, causing rapid corrosion. The development of slag-resistant linings therefore results from using suitable refractories with the appropriate chemical composition.

Increasing the operating temperature invariably accelerates the slag attack, since it reduces the viscosity of the slag and its reaction products, and accelerates the chemical reactions involved.

Like refractories, slags can be acid, basic, or neutral. Basic slags usually contain 50–75% MgO, CaO, FeO and alkalis, while acid slags contain up to 72% $SiO_2$.

An example of a neutral refractory is chromite which resists acid and basic slags roughly to the same degree. Since slag reactions depend on differences in the chemical natures of the reactants, it might be expected that acid linings should resist acid slags and basic linings should resist basic slags. However, in practice the reactions are more complex than this, and any predictions about slag lining reactions should be based on a study of the relevant phase diagrams (*see* Chapter 3), and where possible checked by trials.


**Effect of bonds**

The composition of the bond in the refractories is important. Many refractories are prone to slag attack through their bonds, rather as complete furnace walls are susceptible through their mortar joints. Magnesia particles in magnesite bricks are bonded with monticellite which is the weak part, and through it the slag may enter and bring down the lining. The answer sometimes is to use magnesite with a forsterite or spinel bond. Direct-bonded refractories, that is, those with only one mineral phase, are usually much more slag resistant than multiphase compositions, or at any rate the slag attack is more uniform.

The porosity of the lining is also an important factor in slag action. Slag attack generally rises if the porosity and the pore sizes increase, and the manufacturer's aim in fabricating slag-resistant linings should be to produce a structure with fine pores, regardless of the total porosity, which may be governed by other considerations. In considering pores the investigator of slag is concerned with the sucking action or permeability of the lining, and it is known that at 1 500–1 650°C the pores below 5-$\mu$ diameter cannot suck up most common slags used in metal processing. Another factor is whether the pores are open or closed. Of course, if the slag dissolves the refractory (and first it must wet it) both types of pores will be open to its attack. In order to reduce slag attack the total porosity must be reduced.

The wetting of refractories by slag depends on the compositions and temperatures. Carbon, for example, is not wetted by slags and so slag attack is minimal.

Methods of determining the slag resistance of refractories in laboratory conditions do not accurately reproduce service conditions in furnace linings. Simple methods involve drilling holes in specimens, filling these with powdered slag and melting under various conditions in furnaces, followed by an examination of the sections of the specimens, and possibly some quantitative measurements of the area of slag attack.

More recently methods have been tried which more closely resemble service conditions. For instance, the resistance of refractories can be determined in slag formed in the production of magnesium by the silicothermic method in vacuum electric arc furnaces with a liquid slag bath. The service of refractories in these conditions is greatly complicated by the features of the melting process with a deep slag bath. Special equipment has been designed, consisting of an electric-resistance furnace fitted with graphite elements, graphite crucibles and rods. The slag convection process is reproduced by rotating the specimen of refractory in the molten slag. The method is suitable for temperatures up to 1 600°C.

### Wetting of refractories by molten steel
The manner and extent of the wetting of refractories by molten metals are very important, since before a refractory can be corroded by liquid metal it must be wetted by it. This problem has not yet been thoroughly investigated. Killed carbon steel wets alumina–silicon carbide and zircon refractories to a lesser extent than other refractories used in the steel industry, while these steels readily wet silicon carbide products. Dense and porous alumina refractories are wetted to the same extent by these steels.

### Molten glass resistance
Refractories used in the construction of glass furnaces come into direct contact with molten glasses of various compositions and degrees of corrosiveness. Other conditions in these furnaces are particularly severe (variations in temperature from end to end, causing mechanical stresses and spalling, etc.). However, the main concern in developing refractories for the melters of, say, tank furnaces is with the glass corrosion resistance. The atmosphere inside glass furnaces invariably contains alkali vapours and these may condense on the cold parts of the furnace linings, causing attack. Alkalis penetrate into the pores of the refractory, condense there, and when the temperature rises, fuse and cause rapid corrosion and breakdown from within.

The refractory in the molten glass area, that is, in the tank, is open to the action of highly alkaline silicates and, when sulphate batch is being handled, to direct contact with molten alkali. Sodium sulphate with a melting point of 885°C reacts with lime to form a glass only at a temperature of about 1 440°C, and when sulphate batch is being melted the glass surface invariably carries an alkali skin. This alkali can penetrate all cracks and pores in the glass-tank blocks and is a common cause of breakdown in refractory linings.

The density of tank blocks is a most important factor since the least viscous components of the melt, with low surface tension values, readily wetting the refractory, are sucked into it through the capillaries existing in all types of refractories. Alkalis and highly alkaline silicates are readily sucked up in this way. The materials then react with the amorphous phase of the refractory, and the resulting melt penetrates the entire refractory and dissolves its glassy phase, the crystals of the structure (grog particles) being eroded by the flow of the molten glass and thereby exposing new sections for subsequent corrosion.

Once the glass has been melted other conditions operate within the tank furnace; attack is mainly due to physical factors in the glass stream. These factors include the differences in temperature, which may be considerable from one end of the tank to the other; differences in specific gravity in the glass; and differences in the rate of flow of the glass delivered for working.

The factors governing refractory life in glass furnaces are the properties of the glass; the chemical composition of the refractory; the physical structure of the refractories, mainly porosity; the permeability and wettability of the refractory by the melt, and the sizes and shapes of the refractory grains.

## POROSITY

The porosity of refractories determines other properties such as slag resistance and spalling resistance. The porosity and mechanical strength of a refractory can often be used as quality control factors for the manufacturing process and to evaluate the effect of various production factors on the properties of the finished goods. The pores in a refractory material may be present as a result of deliberately incorporating various additives such as sawdust, cork and other combustibles, and also by using foaming (air-entraining) methods. In the cold state the porosity and strength of refractory materials are closely connected. Increasing the porosity normally reduces the strength. The structure of refractories is determined by the density; the porosity; the apparent density; the surface area of the pores; their distribution and sizes, and whether they are communicating or not. The mechanical strength of refractories is judged from the compressive strength, and in some cases from the bending, torque and tensile strengths.

Most refractories are to some extent porous. The number, sizes and shape of the pores vary considerably and the porosity, for instance of insulation, can be as high as 80%. Normal firebrick has a porosity of about 15–28%, and high density products down to 2–5%. A few specially fabricated materials have zero porosity, that is, the true or theoretical density.

Since the manner in which a slag or molten glass penetrates a refractory is determined by the nature of the porosity, it is important in considering this factor to know whether the pores are communicating or closed. In ordinary refractories most of the pores are communicating, and can be seen on the surface of the products by filling them with a coloured liquid; these pores are open. Some of the pores are isolated and inaccessible to the coloured water; these pores are closed.

It is therefore essential in discussing porosity to differentiate between the true or total porosity of the products, which is made up of the closed and open pore systems, and the apparent or open porosity, which is made up only of open pores. There are standard methods of determining these factors. The pore sizes in refractory materials vary from very large to very small (from fractions of a millimetre to tenths of an Ångström unit). Pore distribution is usually irregular, the large pores being found between the grains of grog and the bond constituent, and the very fine pores in the grains themselves. Much will depend on fabrication methods and on the grain size composition of the batch. The grinding of raw materials and fired raw materials, and the classification into fractions are fundamental processes in refractories production. They are decisive for quality and service life. Grinding also affects certain other important properties, in particular slag and thermal-shock resistance, since it determines the porosity of the fired goods.

The apparent porosity is determined by the volume of liquid which will be absorbed by the pores when the specimen is boiled or kept in vacuum. If the material is water-saturated, the apparent porosity can be calculated from

$$P_A = \frac{d_2 - d_1}{V} \times 100, \%$$

where $P_A$ is the apparent porosity (%); $d_1$ is the weight of the absolutely dry specimen, in grammes; $d_2$ is the weight of the same specimen saturated with water, in grammes; and $V$ is the volume of the specimen ($cm^3$).

The water absorption ($W$) of a refractory is the ratio of the weight of the absorbed water to the weight of the specimen, and equals

$$W = \frac{d_2 - d_1}{d_1} \times 100, \%$$

The ratio between the apparent porosity and the water absorption will therefore equal the apparent density:

$$\gamma_A = \frac{P_A}{W} = \frac{d_1}{V}, g/cm^3$$

The true porosity of a refractory is the ratio of the volume of open and sealed pores to the bulk volume.

The bulk density is the ratio of the mass of the material to its bulk volume (also called apparent specific gravity).

The apparent solid density is the ratio of the mass to the apparent solid volume of a material (also called apparent true specific gravity).

The true density of a refractory is the ratio of the mass of a material to its true volume (also called powder density and absolute density).

### Slag penetration

The method by which a slag, molten metal, or molten glass penetrates into a refractory structure depends on the porosity; the process is complicated. For instance, only pores open at both ends, that is, intercommunicating, will suck up the slag or other liquid. The blind pores do not take part in the transmission of liquids or gases in the structure. The open porosity is therefore divided into permeable and impermeable forms. The permeability is determined not only by the pore geometry but also by the particular properties of the gas or slag.

## PERMEABILITY OF REFRACTORIES

As mentioned above, this will depend on the extent of the communicating, open pores which are available to transmit gases. The sealing of a refractory structure against penetration by gases is important for several reasons. For instance, coke ovens used for the manufacture of coal gas and coke must obviously be gas tight to prevent the loss of gas through the lining. However, in other cases the producer of the refractory deliberately makes the materials gas permeable, as for instance in soaking furnaces, in order to provide uniform temperature distribution and to remove flue gases through the furnace walls and not through other sections of the structure. Gas permeability usually diminishes as the temperature rises.

The magnitude of gas permeability is defined as a coefficient $K$, which signifies the quantity of air or gas in litres passing through a refractory wall with an area of 1 m$^2$ and a wall thickness of 1 m in a period of 1 h, with a pressure difference of 1 mm water. The $K$-value can be determined from the following formula:

$$K = 4\ 586\ 000\ Vh/d^2tp$$

where $K$ is the gas permeability coefficient, $V$ is the volume of transmitted air or gas (litres); $h$ is the height of the specimen; $d$ is the diameter of the specimen (mm); $t$ is the time for air transmission (sec); and $p$ is the working pressure (mm water).

It should be emphasised that porosity itself does not directly determine the gas permeability of refractories, although it is obviously important.

The precise nature of the pores and whether they are open or closed and communicating determines the gas permeability. When a furnace is heated the structure of the material changes, and the gas permeability alters much more markedly than the porosity of the refractory.

## MECHANICAL STRENGTH

The measurable strength of brittle solids such as refractories is found to be much lower than the theoretical strength calculated from the interatomic bond strengths, since in calculating the theoretical strength we start from the ideal, regular structure of the crystal lattice. The theoretical strengths of all forms of ceramic and glass materials are much higher than those realised in practice.

Various types of strength are relevant in refractory materials science (Fig. 2.4). The compressive strength is normally determined as the crushing

FIG. 2.4.   Instron apparatus for measuring strengths of materials. *Photo:* Courtesy British Ceramic Research Association.

strength at room temperatures, and like the density gives some indication of the degree of vitrification or sintering of the products after firing. It therefore depends on the composition, fabrication conditions, firing temperature, and other factors. The mechanical strength of refractory materials is important not only in the furnace structure but also in the carriage of the materials from manufacturing plant to user sites. Standard crushing strength tests are used for determining the property. Bending, rupture and torque factors at room temperature are not of great importance to refractory users. The rupture strength is usually about a fifth to a tenth of that of the compressive strength, and the bending strength a half or a third. However, manufacturers frequently use quality control data on compressive strengths as an indication of the consistency of the manufacturing process for a given type of product.

The abrasion resistance of refractories is important, especially when they are to be used in furnaces that carry abrasive charges, for instance, the blast furnace. Obviously the abrasion resistance will depend largely on the hardness, and on the effectiveness of the bonding of the particles making up the structure of the refractory. The specific hardnesses of the granular material used in the refractories will determine in general the hardness of the product, but attention must also be given to the type of bond. Many refractories have to withstand the abrasive action of dust carried along by furnace gases. The abrasion resistance diminishes if the surface of the refractory lining is softened. In fact, the ultimate failure of a lining may be determined by the abrasion resistance of the refractory.

The impact resistance of refractories is also another important mechanical property, since when furnace charges enter a furnace, such as a blast furnace or lime kiln, the strong impact action may cause the lining to chip or break off.


## CHEMICAL AND MINERAL COMPOSITIONS AND CHANGES IN SERVICE

Apart from the changes already discussed which occur in refractories during service, the user must be aware of important changes in the mineral and chemical compositions during service at high temperatures. These changes may so alter the overall properties of the refractory lining that any information obtained from tests that do not take them into account will fail to give proper indications of the way in which the lining will behave in service.

Many refractories, when acted upon by slags and temperature variations, acquire a zoned structure, and the zones have very different chemical, physical and other properties. Zoned structures commonly develop in forsterite, magnesite and chrome–magnesite products used in electric steel-melting furnaces, open-hearth furnaces, and even in kilns used in the ceramic industry (pottery kilns for instance). Since the properties of the various zones differ from each other, stresses will be produced which may cause spalling and even collapse of the lining. The zoned structures in refractory linings normally result because of the movement of fluxes which enter the refractory from the furnace gases, or move from one part of the refractory structure to another as a result of differences in temperature (temperature gradients). Investigations into the diffusion processes leading to zone formation have established the precise mechanism of failure of refractory materials.

The phase composition and microstructure are studied by using polished and unpolished sections in transmitted and reflected light on specimens taken from the used linings, and also by measuring the refractive indices of various components in the refractories. Microscopic (optical and electron) techniques are also employed to determine the changes occurring in refractories during service.

X-ray diffraction studies are also employed to determine phase compositions. Yet another technique is that involving differential thermal analysis which, when combined with X-ray diffraction data, yields information about the temperatures at which the crystalline phases in the refractory undergo changes during heating. By employing all these techniques, the refractories researcher can obtain an overall picture of the potential performance of a given material in a given lining. However, much remains to be discovered about the behaviour of refractories in service.

## SHAPE AND SIZE TOLERANCE

Precision of shapes and dimensions in refractories is important for several reasons. Joints filled with refractory mortars are normally inferior in refractory properties to the bricks themselves, and therefore any deviation in shape or dimensions will tend to make the joints larger or more irregular, thus exposing the finished lining to greater attack. It is usually the aim of a refractory bricklayer to make the joints as thin as possible, so uniformly sized and shaped refractory bricks make his job easier. The nature of the manufacturing process, especially for bulk refractories, such as common firebrick and silica brick, is such that it is virtually impossible to guarantee that all bricks in a batch will be perfectly shaped and have exactly the same sizes. However, quality control methods in the industry are being constantly improved and fairly close tolerances are now possible. Statistical quality control methods are extensively used in the industry to check products leaving the kilns and to obtain information about quality variations.

## THERMAL CONDUCTIVITY

The capacity of refractory materials to store or conduct heat has an important effect on their behaviour in furnace linings. The property is measured by the coefficient of thermal conductivity, which is the quantity of heat passing through a wall of a standard thickness over a standard area with a difference in temperature at the hot face and the cold face specified by the test. The usual units are kcal, 1 mm thickness, 1 $m^2$ area, and a temperature of 1 degree, so that the conductivity is expressed as

$$\lambda = \text{kcal/m h deg}$$

The passage of heat through refractories determines the amount of heat lost through the refractory linings and also the behaviour of the refractories themselves; for instance, thermal conductivity has an important bearing on the spalling resistance of the brick (*see* page 17).

The thermal conductivity of most refractories increases with temperature. However, certain highly crystalline products show a decrease in thermal conductivity with temperature rise, *e.g.*, magnesite brick, whose conductivity at room temperature is about 5 units, falling to 3 units at 100°C. Other important materials which show this relationship are silicon carbide and, to a lesser extent, alumina. For silicon carbide, the thermal conductivity drops very sharply as the temperature rises.

Thermal conductivity depends on the chemical and mineral compositions of the material, and the structure, particularly the porosity. As the porosity rises, the thermal conductivity diminishes, and this relationship is used in the manufacture of heat-insulating refractories. The relationship between porosity and thermal conductivity is not very clear at very high temperatures. When the pore size of a refractory increases, other conditions being equal, at elevated temperatures the conductivity is markedly increased.

Most refractory products are poor heat conductors. Exceptions are graphite, magnesite, and silicon carbide. For instance, the thermal conductivity coefficient of firebrick and silica is only about a fiftieth of that of the metals. The property of thermal conductivity in refractories is anisotropic, and partly depends on the direction in which the pressing force was applied during fabrication. Another important factor is the method used to determine the thermal conductivity, a property which is difficult to determine precisely. It is normal to specify that the conductivity be determined in two mutually perpendicular directions in the brick, one of which must coincide with the pressing direction.

Specific heat is the quantity of heat used in warming up unit weight of a material by a given amount. The specific heats of refractory materials are normally expressed in units of specific heat at constant pressure.

Specific heat values are measured in the same way as for other materials and are used in heat calculations and in furnace design.

## ELECTRICAL PROPERTIES

At low temperatures most refractories are dielectrics, that is, they will withstand an electrical stress, but during heating their electrical conductivity rises in most cases, and above 950°C they are quite good conductors. Many refractory materials are used in fabricating heating elements, *e.g.* molybdenum disilicide, silicon carbide, etc., and their electrical properties have been thoroughly investigated. Electric melting and heat-treatment furnaces also require the refractory producer and user to have a sound knowledge of the electrical properties of refractories.

The impurity concentration of the raw materials employed in manufacturing refractories has an important bearing on the electrical resistance, since these impurities reduce the temperature at which a liquid phase appears, which alters the electrical factors. The physicochemical structure of the impurities and their distribution are also critical in their effect on the resistance of refractories.

Information on the most important electrical properties of refractories is normally given in textbooks on electric insulating ceramics under such headings as: dielectric permeability, the temperature coefficient of dielectric permeability, the specific volume and surface resistances, dielectric loss, and electrical or breakdown strength. Naturally the electrical properties of refractories and ceramics are closely dependent on the composition and structure of crystalline phases making up the particular type of ceramic, the composition of the vitreous substance, and the ratio of crystal and vitreous phases in the ceramic. The crystal phases of refractories consist generally of crystals with ionic bonds. Usually there are no free electrons in refractory materials, in contrast to metals.

A wide range of special refractory oxide and non-oxide compounds which have special electrical and other properties is being produced. These include carbides, nitrides, silicides, borides, and cermets (combinations of metals and ceramics). Their production is a specialised branch of the ceramics and refractories industry. Further information on their properties will be found in Chapter 17.

## VOLATILISATION

Most solid oxides volatilise with rise in temperature. The effect depends on the type of crystal lattice and its energy, and is therefore dependent on the strength of the bond between the atoms or the ions in the structure.

FIG. 2.5. Preparing specimens for heating under low pressure. Volatilisation of certain refractory constituents may occur under such conditions, weakening the structure and causing premature failure. Low-pressure technology is of increasing importance today. *Photo:* Courtesy British Ceramic Research Association.

The volatilisation of refractory constituents from certain types of refractory structures is very important, not only because of the resulting changes in the structure, but also because of the possible effects of contamination of the materials being processed in the furnace (*see* Fig. 2.5). The relevant factors are temperature, vapour pressure and surface area. Magnesite refractories are very prone to volatilisation (q.v.).

*Chapter 3*

# The Raw Materials of Refractories

Any appreciation of the behaviour of heat-resistant materials in service must be based on an understanding of what distinguishes a refractory from other materials. The chemical elements, compounds and systems of compounds employed to make refractories must therefore be discussed.

The list of raw materials used to make refractories is very extensive and it is being continuously revised and enlarged. Refractory raw materials can roughly be divided into acidic (*e.g.* silica and siliceous); basic (*e.g.*

Fig. 3.1.   The inside of a Pugsealer for a two-stage extruder for processing plastic refractory raw materials. De-airing and compaction of clays and clay-grog mixtures are important processes in producing high-grade bricks. *Photo:* Courtesy Thos. C. Fawcett Limited.

magnesite and lime); and neutral (*e.g.* chromite). Fireclays are perhaps the commonest refractory raw materials, although they are not now as important in the raw materials economy as they used to be. These may be bauxitic, diaspore-types, flint-clay, or kaolinitic. Non-plastic minerals

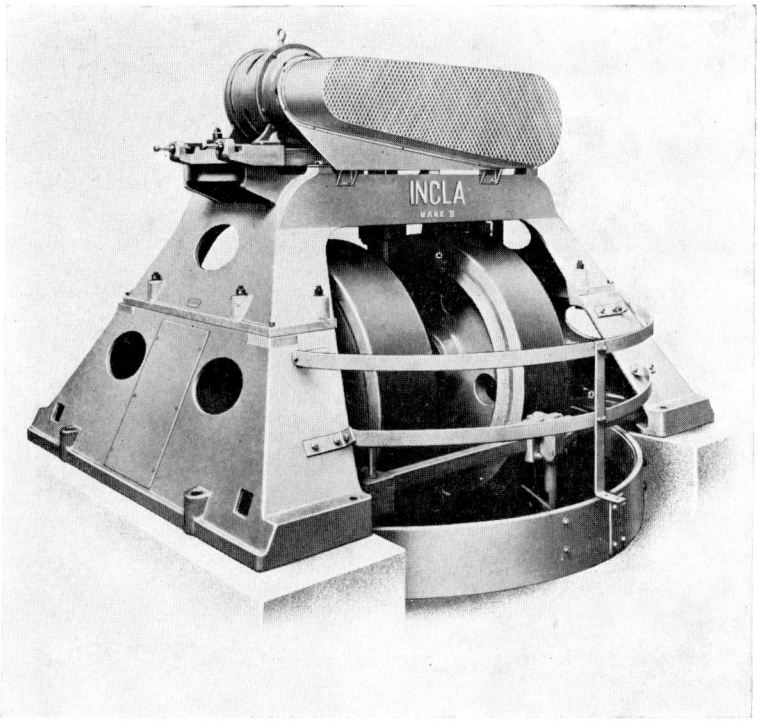

FIG. 3.2.    The Mark 2 Incla grinding mill for dry grinding shales and friable refractory materials. *Photo:* Courtesy Thos. C. Fawcett Limited.

containing varying quantities of alumina are also important and fairly widespread. They include corundum, andalusite, kyanite, sillimanite, natural hydrated alumina, bauxites, sialites, and alites.

The silica and siliceous materials include gannister, firestone, diatomaceous earth and various sands and sandstones. The magnesia and lime materials include dolomite, olivine, forsterite, dunites, serpentines, talc and talc stone and magnesites and dolomites.

Other refractory materials are based on chromites, zirconium compounds, carbon, and rare-earth oxides, as well as titanium dioxide, thorium, beryllium, and uranium oxides.

FIG. 3.3. The inside of the revolving pan of an Incla mill for grinding refractory raw materials. The perforated high-tensile steel grids have tapered holes, through which the crushed particles pass on to conveyors for further processing. *Photo:* Courtesy Thos. C. Fawcett Limited.

FIG. 3.4. A rotary nylon-mesh screen for classifying refractory raw materials after grinding. It is inclined 5° and fed by a dry grinding mill. Coarse grains are returned to the mill. *Photo:* Courtesy Thos. C. Fawcett Limited.

These materials and combinations of them are quarried and used in the natural state or subjected to various degrees of purification by processing (Figs. 3.1–3.4). They are then employed to manufacture a very wide range of refractory products, ranging from the common firebrick to highly sophisticated components for use in nuclear reactors, space craft and missiles.

## MELTING POINTS

Substances with melting points or fusing temperatures above 1 580°C are normally used for making refractories. Raw materials with lower fusing points are sometimes processed to remove fluxing impurities and to yield a product meeting the requirements of the refractories industry.

### Compounds melting above 1 580°C

In the Periodic System of the chemical elements there are only nine elements with melting points of 1 000–1 500°C, there are 15 elements with melting points of 1 500–2 000°C, and 10 elements with melting points above 2 000°C. In any refractory system, increasing the number of elements tends to lower the maximum fusion point of the compounds involved, and the probability of forming a refractory material from two simple oxides is almost twice as great as that of forming such a material from three oxides. Accordingly, refractories technologists search for new oxygen-containing refractories mainly in binary systems of simple oxides. Of the 9 200 binary compounds that can possibly be formulated out of the 22 refractory oxides, only 1 010 are refractory, that is, 11 %. The probable number of ternary compounds is about 148 000, and most of these have not yet been obtained. Statistical methods suggest that the probable number of refractories among them is about 2 870, which is only about 2 %.

For the quaternary compounds the number of refractories is of the order of several hundreds, or about 0·01 %. Not a single refractory is known containing five elements if we ignore solid solutions.

The highest melting point of any refractory is possessed by graphitic carbon, which in an argon atmosphere at a pressure of 150 atm melts at 4 200°C. However, at normal atmospheric pressures in an inert atmosphere, carbon volatilises at 3 500°C. In oxygen or air it burns long before this stage is reached.

Other highly refractory materials are tungsten (with a melting point of 3 370°C), rhenium (3 000°C), osmium (2 700°C) and molybdenum (2 620°C) and the disilicides, of which molybdenum disilicide is used for fabricating special refractories and electrical heating elements.

All the refractory materials with melting points about 2 000°C, except carbon, are classified as rare elements, and are used only in very small amounts compared with bulk refractories such as calcined alumina, fireclay, magnesite, etc. The majority of refractory materials are compounds or complicated systems of compounds such as oxides, aluminates, etc. Only carbon, as an elemental refractory material, is used for mass production.

The Periodic System of chemical elements can be used to give an indication of the manner in which the melting points of the elements alter in terms of their chemical properties. The elements with high melting points crystallise into dense structures such as cubic and hexagonal. Of the chemical compounds the highest refractoriness is possessed by nitrides, carbides and some oxides. The siliceous materials and other acid refractories such as titania have the lowest fusing points of the refractory materials, while the basic oxides such as MgO and CaO have the highest fusion points. The amphoteric or neutral oxides, such as alumina and chromium oxides, take an intermediate position. The vast majority of refractory materials are based on chemically basic oxides. However, a very important section of refractory materials is taken up by silicates and aluminates, phosphates, spinels, and aluminosilicates. It is reckoned that such compounds number about 4 000 (melting point greater than 1 853°K), but only about 500–600 have been studied.

## DISTRIBUTION IN NATURE

In the earth's crust the oxygen compounds of silicon, aluminium, iron, also rare earths and alkali metals such as calcium, magnesium, sodium and potassium, are very widely distributed. The nature of this distribution, of course, determines the economic viability of using any material as a refractory. At the time of writing the oxides of silicon, aluminium, magnesium, chromium, calcium, beryllium and zirconium, and compounds and mixtures of these oxides, constitute, together with carbon and graphite, the basis of the raw materials store for the refractories industry. Man-made compounds such as carbides and nitrides are also becoming increasingly important.

## PHASE DIAGRAMS IN REFRACTORIES

The phase diagram is a graphical device employed by the inorganic chemist to study the possible equilibrium states of a system in various conditions. Phase diagrams can be used to answer all questions arising in a study of heterogeneous equilibria, but it must be remembered that the information obtained refers only to systems that are in equilibrium and since most of the refractories used in actual service conditions are not in this state, allowances must be made. A common approach by the refractories technologist is to select two or three components found in the refractory material and constituting a large part of the compositions. In this way the problem is simplified and it is only necessary to investigate the binary or ternary system (two or three components).

The following information can be obtained from a study of phase diagrams:

1. The mineral components of the materials at any temperature.
2. The temperature at which liquid first forms.
3. The alterations in the concentrations of liquid and the liquid compositions with temperature.
4. The order of crystallisation of the phases during cooling and their compositions and amounts.
5. The chemical solubility of components or phases at different temperatures.

Information used in constructing phase diagrams of interest to refractory-material technologists is obtained by the following methods:

(a) By constructing cooling or heating curves from which clear thermal effects, such as breaks or bends in the curve, are noted, indicating phase changes in the system, such as crystallisation, polymorphic inversions, or the formation of chemical compounds.

(b)  By chilling the system, followed by determination of the mineralogical composition, using such techniques as petrographic or X-ray methods. In the latter case sudden cooling fixes the crystalline phases, and the liquid phase is changed into the particular glassy phase existing at the stated temperature.

In drawing up these phase diagrams for silicate and other refractory systems it is usual to concentrate on the second method, since the first is not very suitable for refractories because of undercooling.

Apart from the fact that all data 'read off' phase diagrams applies only to equilibrium conditions, they do not provide information about the viscosity of the melt, information which is of the utmost importance in studying the behaviour of refractories. Furthermore, chemical solubility considerations do not take account of porosity or capilliary action producing changes in the composition as the fluxing elements move across temperature gradients. The comprehensive collection of phase diagrams of interest to ceramists and users of refractories published by the American Ceramic Society (Phase Diagrams For Ceramics) compiled by Levin, McMurdie and Robbins, is the standard work on the subject, and may be consulted for the most recent information.

The following descriptions of the various refractory raw materials are intended to give some indication of the range of materials available and some of their properties. However, the amount of information available on each one of these materials is now considerable, and primary sources should be consulted for more detailed information.

## FIRECLAYS

Fireclays constitute a main category of refractory raw materials and are used for making firebrick and siliceous refractories for various purposes. They are made in a wide range, and the quality is normally related to the alumina, $Al_2O_3$, content. Fireclays have varying degrees of plasticity (Fig. 3.2) and may contain varying quantities of residual minerals and impurities. During firing they lose their chemically bonded water and plasticity and upon the attainment of a certain temperature acquire mechanical strength.

In Britain fireclays are usually found with coal measures and consist chiefly of kaolinite and free silica. Extensive deposits are found in Scotland (Ayrshire) and in the Midlands. Most countries of the world have some fireclays and the leading industrialised countries such as the USA, USSR, Australia, and European countries possess adequate supplies of common and high-grade fireclays. Useful minerals in fireclays, in addition to kaolinite, may be halloysite and monothermite (illite, and hydrated muscovite) and also quartz, mica and hydrated mica. The contaminating and undesirable minerals consist of limonite, pyrites, limestone (calcium carbonate and also ferrous carbonate) and some organic compounds. The dark coloured fireclays owe their colour to the presence of finely distributed organic matter, which of course burns out upon firing. The organic matter may have a considerable influence on the plasticity and other rheological properties of fireclays.

Various classifications are used for fireclays, depending on their alumina concentrations. In some countries clays containing more than $30\%$ $Al_2O_3$ are called basic clays, those containing $15-30\%$ alumina are called semi-acid or siliceous clays, and those containing less than $15\%$ $Al_2O_3$ are called acid clays. The ratio of the alumina to silica usually indicates the type of clay, and if this ratio is roughly $1:2$ the clay is normally of the kaolinitic type.

The refractory characteristics of fireclays depend mainly on the mineral composition. The highest PCE values are recorded for pure kaolinite clays ($1\,760-1\,770°C$), with any impurities, such as quartz and iron, tending to reduce the PCE value.

Although the importance of fireclays as refractory materials is still considerable, there is no doubt that their place in furnace construction is gradually being eroded by the need for materials that can be processed and shaped with a degree of consistency that is not possible with clay. The economics of furnace and kiln design and construction are also militating against firebricks and fireclay-based refractories, because it is generally more economical to use more expensive and longer lasting refractories which tend to reduce the costs of construction and boost yield of metal, etc., per ton of refractory.

## KAOLINS (CHINA CLAY)

Kaolins which are based on the mineral kaolinite, $Al_2O_3.2SiO_2.2H_2O$, are widely used for making high quality aluminosilicate refractories. Primary kaolins are formed by the weathering of igneous, metamorphic and sedimentary rocks containing feldspar and mica (for example, granites). The largest and purest deposits of kaolin are to be found in Cornwall, England. Other important deposits are found in Czechoslovakia and the USA. The Soviet Union also possesses some 70 deposits of kaolins of variable quality. Kaolins are distinguished from kaolinitic fireclays by the relatively greater coarseness of their particles, the much lower concentrations of impurities and the lower plasticity (*see* Fig. 3.5).

The properties and compositions of kaolins in various deposits are largely governed by the conditions under which they were formed. Residual kaolins (primary) have a rather coarse grain size composition with particles mainly of 0·1–0·005 mm in diameter, and often contain coarse-grained residual quartz and undecomposed fragments of the original rock. They therefore have to be beneficiated in order to produce commercial china clay. Secondary kaolins are similar in grain-size characteristics to some fireclays and ball clays and have much higher concentrations of very fine particles (up to 80% particles <0·005–0·001 mm).

FIG. 3.5.   This photograph, taken with the electron microscope, shows the flat, plate-like shape of particles of kaolin. *Photo:* Courtesy English China Clays Group.

When heated, kaolins will start to sinter or vitrify in the temperature range 1 450–1 500°C. Their PCE is normally 1 700–1 780°C, when they are free of fluxing impurities. China clays are used for manufacturing high-alumina refractories, as high grade bonding clays, and also (when calcined) as china clay grog (sold in Britain under the name molochite). This kaolin chamotte or molochite is used to make a range of refractories for use in blast furnaces, in kiln furniture and refractory cements and concretes. The china clay is normally calcined in rotary kilns and then ground to produce size-graded materials.

## SILICA MINERALS

These include sandstones, quartzite, gannister, firestone and diatomaceous earth. They contain varying quantities of silica, $SiO_2$ and impurities. The main raw materials, for instance, for manufacturing silica bricks (dinas) are quartzites, sandstones and quartz sand. These materials are found extensively throughout the world and are processed or used in the natural state to make a range of acid (siliceous) refractory materials. The forms of silica and the changes which occur when silica is heated are of fundamental importance to the producer and user of silica brick; these aspects of the subject are discussed in Chapter 8.

The chemical composition of most common quartzitic materials used for making silica brick is 92–98% $SiO_2$; 0·5–2·5% $Al_2O_3$; 0·15–3·0% $Fe_2O_3$; 0·2–1·6% CaO; 0·1–0·6% MgO; 0·2–0·5% alkalis; and loss on ignition 0·2–1·0%.

The amounts of alumina in quartzites used for making silica brick are of the utmost importance, since by reducing the alumina content of the fired brick it is possible to improve considerably the performance of the silica brick in furnace linings (*see* Chapter 8).

## ALUMINA RAW MATERIALS

These constitute one of the most important categories of refractory raw materials. Materials of interest in refractories technology containing alumina range from common fireclays to superduty alumina refractories consisting of almost 100% $Al_2O_3$. The range includes corundum, andalusite, sillimanite, kyanite, (cyanite) diaspore, boehmite, and bauxite. Pyrophyllite, a natural hydrated aluminium silicate with the formula $Al_2O_3.4SiO_2.H_2O$, can also be used to manufacture refractory materials, specifically for ladle brick. Another mineral of high alumina content is dumortierite, with the formula $8Al_2O_3.6SiO_2.B_2O_3$, it is found in the USA, and also as an accessory mineral with andalusite and more rarely in other minerals of the kyanite group.

Corundum is the only form of alumina that remains stable when fired above 1 000°C. It is also known as $\alpha\text{-}Al_2O_3$ and has a fusing point of 2 050°C. It is normally extracted from bauxite and calcined. The chemical composition of corundum corresponds to the theoretical formula of $Al_2O_3$. In nature it is found pure and also contaminated. The refractoriness as measured by the PCE for various deposits of corundum varies from 1 850 to 2 030°C. The purest forms of commercially available corundum (as raw material) contain 95–98% $Al_2O_3$.

Kyanite, sillimanite and andalusite are modifications of a single aluminosilicate with the general formula $Al_2O_3.SiO_2$ and consequently have roughly the same chemical composition but different structures. They are used for making high-duty aluminous refractories for a wide range of purposes. When heated they exhibit varying volume changes, dissociating into mullite and cristobalite, according to the following formula:

$$3(Al_2O_3.SiO_2) = 3Al_2O_3.2SiO_2 + SiO_2$$

It is thought kyanite changes into mullite at much lower temperatures (1 300–1 350°C) with a large volume increase (16–18%). Sillimanite is reckoned to be the most stable of the trio, dissociating at about 1 550°C with an increase in volume of about 7%. Andalusite occupies an intermediate position (3–5%).

Dumortierite is an accessory mineral to andalusite and other minerals in the group. It also contains boric oxide and must therefore be treated with great care as a refractory material. It contains up to 64% alumina, 30% $SiO_2$ and 6% $B_2O_3$ and when heated to 1 500°C loses its boric oxide and water. Complete dissociation occurs, resulting in the formation of mullite and siliceous glass. Dumortierite is not of great significance because of the limited deposits throughout the world.

Hydrated alumina is found commonly in nature and is a very important type of refractory raw material. The group contains mono- and tri-hydrated alumina in the form of diaspore, boehmite and gibbsite (hydrargillate). Diaspore, with the formula $Al_2O_3.H_2O$, contains about 85% $Al_2O_3$, the rest being water. It crystallises in the rhombic form and is very brittle, with a specific gravity of 3·3–3·5. When heated to about 400°C it loses its water of hydration and upon further heating transforms into $\alpha$-alumina (corundum) when its specific gravity is raised to 4·0. The PCE of pure diaspore is 2 030–2 050°C but natural diaspore is contaminated with iron oxides and silica and has a refractoriness of about 1 900–1 940°C. Diaspore materials or concentrates of them need precalcining before use as refractories. Large deposits of stone-like diaspore clays are found in China and the USA and in some parts of the Soviet Union.

Boehmite is identical in chemical composition to diaspore but has a different structure: it is rhombic. The hardness is 3·5 and the specific gravity 3·0–3·06 g/cm³. When heated it changes into $\gamma$-$Al_2O_3$ (cubic form).

Hydrargillite or gibbsite has the formula $Al_2O_3.3H_2O$, with a theoretical alumina content of 65·4%; it is monoclinic with a specific gravity of 2·43, and upon heating in the range 160–230°C it changes to boehmite by losing two molecules of water and subsequently at 370–490°C $\gamma$-$Al_2O_3$ is formed; the $\alpha$-alumina form also starts to crystallise. It is rarely found in the pure form in nature but is widely distributed in many bauxite deposits and is found in fireclays and kaolins as alumina-rich veins.

Bauxites and alites are other important alumina raw materials for refractories manufacture. The chemical composition of bauxite minerals can be either alitic with an $Al_2O_3:SiO_2$ ratio greater than 1, or sialitic with a ratio of less than 1. However, the industrial use of these compounds mainly rests with the bauxites themselves which are alitic and have a ratio $Al_2O_3:SiO_2$ of about 3 or above. Bauxitic raw minerals are extremely variable in composition and depending on the impurity concentrations the specific gravity varies from 2·29 to 3·05; after calcination it rises to about 3–3·5. The purer forms of commercial bauxites have a PCE of about 1 800–1 900°C. The usual procedure with bauxites is to calcine them to convert them into chamotte, a process that is accompanied by a high shrinkage, following dehydration of the alumina (see, however, Chapter 7).

Commercially processed alumina is available in a wide range of grades. Many of the natural forms of alumina discussed above are mined, concentrated and otherwise processed and often calcined to produce specially graded alumina grains for subsequent fabrication into a wide variety of refractories. Sophisticated techniques such as the production of 'bubble' alumina are used for making special materials. Bubble alumina (made by gassing fused alumina) is used in the manufacture of highly refractory insulating materials.

The most important types of refractory raw materials are discussed in greater detail under the sections dealing with the various types of refractory products; for instance, the sillimanite group is discussed under high-alumina refractories (Section II, Chapter 7).

## MAGNESITE

The raw materials for producing magnesite refractories are the natural mineral magnesite, and the chemically produced magnesite, extracted from seawater, which can be processed to obtain a material containing up to 98·5% MgO. Seawater magnesite in countries such as Britain, America and Japan, which do not possess adequate mineral reserves, is a very important source of magnesite for refractories. Briefly the process is to treat the brine with calcined slaked dolomite or lime. The hydrated magnesia is precipitated and can be filtered off and calcined for use as it is. The Soviet Union has vast deposits of magnesite minerals but is also developing the seawater process. The mineral magnesite ($MgCO_3$) consists almost totally of the crystalline form of magnesium carbonate. The pure mineral contains 47·6% MgO and 52·4% $CO_2$. It is white or slightly off-white with a glassy sheen, has a hardness of 4–4·5 and a density of 2·0–3·1. Magnesite forms a continuous series of solid solutions with siderite, $FeCO_3$. It forms a double compound known as dolomite with $CaCO_3$ (see below).

Magnesite refractories are very important in many industries and especially in steel melting. Magnesite has a fusing point of about 2 500°C and other important 'refractory' properties. The impurities of magnesite, which include iron, manganese and calcium, have a critical effect on its refractory properties.

Calcium oxide, which after burning of the magnesite prior to use is in the free state and likely to absorb moisture, forming brittle hydrated calcium oxide hydrate, is produced from calcium silicates such as monticellite. It is very prone to polymorphic changes, which tend to break up the material into powder. This behaviour is obviously undesirable in the manufacture of refractory structures. Other impurities such as silica, alumina and $Fe_2O_3$ are harmful because they reduce the refractoriness of the magnesite, so it is necessary to keep them to a minimum, particularly for high-duty products. The calcium oxide content of the best magnesite should be less than 2%, silica not more than 3%, and the total MgO at least 92%. For so-called fettling magnesite, the MgO content is about 85%. The iron content is not normally specified but it should not be more than 5%. Magnesite, $MgCO_3$, decomposes most intensely when heated to 640–800°C, according to the reaction

$$MgCO_3 = MgO + CO_2$$

Low porosity magnesite is obtained by dead burning seawater material or the rock. This 'total' firing process eliminates all the water and also tends to eliminate shrinkage, producing a dense material for refractories use. The impurities of the natural mineral help to form liquids during the dead-burning process, and these help to reduce the porosity by vitrification.

Small quantities of silica also form a forsterite bond which increases the hot strength under load of refractories made with these materials. However, with higher contents of silica and with a calcium oxide–silica ratio of less than 2, fusible compounds in the form of monticellite and merwinite are formed. The behaviour of briquetted magnesite in firing is discussed under magnesite refractories (*see* Section II, Chapter 9).

Subsequent heating of the magnesite from 800°C to about 1 400°C sees the commencement of sintering, leading to the conversion of the magnesium oxide into periclase (MgO), which is the most stable crystalline form of calcined magnesite. Periclase has a melting point of 2 800°C and a specific gravity of 3·58. In solid solution with traces of FeO it is the chief ingredient of magnesite refractories. Large single crystals of periclase have a noticeable ductility—an important point in the use of these refractories. The quality of the sintered magnesite used as a raw material is largely determined by the quantity and degree of periclase crystallisation. The magnesioferrite developed during calcination acts as mineraliser. When the magnesite has been burnt to these temperatures it is then known as dead-burnt, sintered magnesite, or metallurgical powder. Thermal sintering is accompanied by marked shrinkage as mentioned above and an increase in specific gravity to 3·55–3·58, compared with 3·07–3·22 for poorly calcined raw material. It should be noted that the true specific gravity does not alter with increasing calcination temperature; the increase in specific gravity is a result of vitrification and shrinkage. The following reactions occur during the calcination of natural magnesite containing the usual impurities:

1. $MgCO_3 \rightarrow MgO + CO_2$
2. $3FeCO_3 \rightarrow Fe_3O_4 + 2CO_2 + CO$ } First stage of decarbonisation up to 800°C
3. $CaCO_3 \rightarrow CaO + CO_2$
4. $MgO \rightarrow periclase$
5. $2Fe_3O_4 + 3O + 3MgO \rightarrow 3(MgO, Fe_2O_3)$
6. Series of solid solutions as follows
   $nMgO + mFeO \rightarrow (nMgOmFeO)$
7. $MgO + SiO_2 \rightarrow MgOSiO_3$;
   $2MgO + SiO_2 \rightarrow Mg_2SiO_4$

} Second stage 800– 1 400°C; commencement of sintering

The third stage, taking place at 1 400–1 600°C, results in the termination of sintering with a growth in the periclase crystals, and the formation of forsterite and monticellite.

Other forms of magnesite used in the industry are those known as brucite, $Mg(OH)_2$, and also synthetic magnesite, known as heavy magnesia and light magnesia. The last two are basic magnesium carbonates which contain water of hydration, and also need calcination before use in refractories.

It is important to differentiate between so-called caustic magnesite (that burnt at 800–1 000°C), and sintered or dead-burnt magnesite (1 500–1 600°C). The former will react at room temperature with moisture, that is, it is hydrated, with the evolution of heat. The material exhibits an increase in volume (almost double), so that caustic magnesite cannot be used as the main constituent of batches for producing magnesite products. By bonding caustic magnesite with aqueous solutions of such salts as magnesium sulphate, calcium chloride, magnesium chloride and zinc chloride, it is possible to set it so as to form a strong, stone-like material which can be used as an unfired refractoiy and in refractory concretes. The rate and degree of hydration of magnesia diminishes with rise in the firing temperature. Even the dead-burnt material will hydrate when reheated to 60–80°C and if this material is slaked with solutions of the above compounds it sets. Such a material is sometimes called periclase cement. The variations in the hydration properties of magnesite as a function of burning temperature are due to the different crystal sizes and the different degrees of defectiveness of their lattices.

## DOLOMITE

Dolomite consists of a crystalline rock composed almost completely of the mineral of the same name, a double carbonate of calcium and magnesium with the formula $CaMg(CO_3)_2$. The theoretical composition of dolomite is $30.4\%$ CaO, $27.9\%$ MgO and $47.7\%$ $CO_2$. The impurities are silica, iron oxide, alumina, and manganese oxide, and, less frequently, alkalis. Raw dolomite as used in the steel industry, for instance, normally contains at least $18\%$ MgO and not more than $3.5-6\%$ $SiO_2$. Burnt metallurgical dolomite is of the highest quality; it will be discussed under Dolomitic Refractories.

The crystal form of dolomite is trigonal, its hardness $3.5-4$, and specific gravity $1.8-2.9$. Dolomites are very common in nature, most of them being bonded with carbonates. The mineral impurities to be found in dolomite are quartz, calcite, magnesite, gypsum, clay substance, iron oxides, and various other materials as well as organic substances. Depending on the impurities, dolomites have different colours ranging from grey to reddish. The suitability of dolomites for the refractories industry is governed by their chemical and mineral compositions, and the structure and impurity concentrations.

When fired at $700-900°C$ dolomite dissociates into a mixture of calcium oxide, magnesium oxide, and carbon dioxide; the mixture is called caustic dolomite and is used as a cementing substance. The high-temperature burnt dolomite ($1\ 500-1\ 700°C$), like the magnesite, is called dead-burnt and is relatively resistant to hydration. It has the required mechanical strength and density for use as a refractory material. At high temperatures the magnesium oxide from the dolomite changes into periclase and the calcium oxide crystallises. The sintering temperature of dolomite depends on the impurity contents and the purer dolomites are naturally the most difficult to sinter. Silicates, aluminates and ferrites form from the impurities; this tends to depress the refractoriness of the burnt dolomite but the impurities play an important role as stabilisers which prevent hydration, breaking up and weakening of the products.

Very high-temperature firing of dolomite confers on it a stability which is high enough for it to be transported. However, material fired at lower temperatures has a high content of quicklime which will hydrate fairly rapidly.

To stabilise burnt dolomite for use as a refractory material, the calcined grains are coated with tar to prevent access of moisture from the atmosphere. Tarred dolomite refractories are discussed under this heading. In calcining dolomite to obtain a refractory material it is possible to boost the sintering rate by adding crystalline calcium silicates and other compounds which participate in the formation of the clinker and after firing are converted into the highly dispersed state.

Siliceous serpentine is sometimes used for stabilising dolomites. Providing the proportions are correct the free lime in the dolomite will combine with the silica to form free $CaO.SiO_2$. This increases the stability of the product. The composition of stabilised dolomites is roughly $16.5\%$ $SiO_2$, $1.2\%$ $Al_2O_3$, $2.7\%$ $Fe_2O_3$, $0.75\%$ $Mn_2O_3$, $46\%$ CaO, $30\%$ MgO, and $1.2\%$ $P_2O_5$. Phosphates are added to the raw material to aid stabilisation. The product has a refractoriness of 1 770–1 780°C, compressive strength of 1 000–1 350 kg/cm$^2$ and a refractoriness under load of 1 550–1 610°C with ultimate failure at about 1 680°C. Dolomite refractories made with such materials have lasted for more than 300 days in the back walls of open hearth furnaces and for up to 60 heats in the walls of electric steel melting furnaces.

## FORSTERITE MATERIALS

Forsterite occurs naturally in association with fayalite, and in the mineral olivine. It is a magnesium orthosilicate with a melting point of about 1 890°C. Forsterite refractories are industrially important and mention must be made of raw materials such as olivine or serpentine, as well as the dead-burnt magnesite already discussed, which are used to manufacture forsteritic refractories. The forsterite refractories that find common use in the industry consist mainly of the synthetic mineral forsterite (up to $85\%$) and magnesioferrite (up to $15\%$). The other mineral impurities are present in small quantities and have only a slight effect on their properties. Raw materials for making forsterite products are olivine serpentines, dunite (which consists of olivine at the stage of serpentinisation, containing up to $60\%$ olivinite and $45\%$ serpentine) and talc. Some unusual impurities occur in these minerals, including NiO, MnO, $Cr_2O_3$, as well as the usual calcium, alumina, etc. The calcium and alumina are particularly damaging to the properties of these refractories, since fusible monticellite is formed, resulting in the eventual formation of cordierite with the formula $2MgO.2Al_2O_3.5SiO_2$.

*Olivines* with the formula $(Mg\ Fe)_2SiO_4$ consist of iron varieties of forsterite, or more accurately a solid solution of ferrous orthosilicate (fayalite) in forsterite $(MgSiO_4)$. The materials also contain $2$–$15\%$ impurities, mainly in the form of enstatite, magnetite, talc and sometimes chrome-spinel. The usual chemical composition range of olivine is $45$–$50\%$ MgO, $8$–$12\%$ FeO (more rarely up to $20\%$) together with cobalt, manganese and nickel impurities. The hardness is $6.5$–$7$ and the specific gravity $3.3$–$3.5$ (increasing with increase in the FeO content). The refractoriness (PCE) varies from 1 750 to 1 840°C.

A typical feature of olivines is their capacity for forming a skeleton of crystalline forsterite at 1 500–1 700°C. The gaps between the crystals are filled with the liquid phase which sometimes is as much as 40% of the weight. The mineral forsterite is itself very highly refractory.

*Dunite* is a magnesium-silicate igneous rock containing olivine and serpentine. The content of these minerals in dunites varies and so the properties alter considerably, depending on the degree of serpentine formation. As the olivine content increases its refractoriness is increased and may reach up to 1 800°C. Increasing the sintering temperature reduces the shrinkage.

*Serpentines* usually contain less than 20% olivine. The main feature of serpentines is high loss upon ignition owing to water of hydration. The refractoriness is 1 500–1 570°C; the specific gravity 2·65. When heated to about 1 000°C serpentine loses its water of hydration and is converted into magnesium orthosilicate. Further heating causes the free silica to combine with the orthosilicate to form magnesium metasilicate with the formula $2MgO.SiO_3$ (clinoenstatite). The metasilicate may also take another form (enstatite), which upon heating above 1 150°C changes irreversibly into clinoenstatite. Since magnesium metasilicate fuses at about 1 580°C it cannot be described as a refractory material, so magnesia is added during processing to make it so. The reaction is:

$$MgSiO_3 + MgO = Mg_2SiO_4 \text{ (forsterite)}$$

The conversion is completed in the solid phase at about 1 450°C, accompanied by densification of the material and an increase in the specific gravity.

*Talc and soapstone.* Talc is a common ceramic material with the chemical composition 31·8% $MgO$, 63·6% $SiO_2$, 4·8% $H_2O$. It fuses at 1 500–1 550°C and is mainly used in ceramics in the form of steatite. If magnesite is present in talc minerals the refractoriness is considerably enhanced; it is then known as talcstone or talc magnesite. Talc minerals are not very satisfactory refractory raw materials and in any case usually require precalcination to completely dehydrate them. It is usual to add magnesite to convert them into forsterite refractories.

The best raw materials for producing forsterite refractories are olivines which contain no chemically bonded water, have a low shrinkage and do not break up during calcination.

## CHROMITE (IRON CHROMITE)

Care must be taken to differentiate between the compound chromite, $FeO.Cr_2O_3$, and chrome ore which is (erroneously) also termed chromite, actually a complicated mineral of the composition $(Mg, Fe).(Cr, Al)_2O_3$

which is present (in amounts up to 80%) in the ore used to make refractories. The important minerals in chromitic refractories are the chrome spinels. In composition they range from chromite $FeO.Cr_2O_3$ (which is rarely encountered in the pure form), to aluminochromite, magnesiochromite, and chromepicotite with the formula $(Mg, Fe)(Cr, Al)_2O_4$. Chromite, $FeCr_2O_4$, melts at 2 180°C. The mineral serpentine, often present as an impurity, has a very adverse effect on the refractoriness of chromite. Olivine and chromitic chlorites also reduce the refractoriness. Pure chromite when heated undergoes no important changes and retains its original volume up to 1 700°C, showing moderate shrinkage above this temperature. Chrome ores containing at least 33% $Cr_2O_3$ are selected for refractories.

Chrome ores are used to make chrome–magnesite and magnesite–chromite refractories. These are obtained from chromite and dead-burnt magnesite with the accompanying forsterite, monticellite, magnetite and magnesioferrite minerals. Chrome–magnesite refractories normally contain 15–35% $Cr_2O_3$ and 42–50% MgO; and the magnesite–chromite refractories 55–65% MgO and 8–18% $Cr_2O_3$. The subdivision is rather tentative, and frequently all combinations of chromite and magnesite in the refractories industry are called chrome–magnesite refractories.

Magnesite–chromite refractories have to a great extent replaced silica in the roofs of open-hearth furnaces with a consequent increase in steel output and faster melting rates. The roof temperatures can be raised by about 100°C, and this facilitates the melting of higher grade alloys. The use of magnesite–chromite roofs also tends to prolong the furnace campaigns.

Further information about the constitution and properties of chromitic refractories will be found under the section dealing with the types of refractories and their applications (*see* Sections II and III).

## GRAPHITE AND CARBON MATERIALS

Three classes of carbon-containing refractories are based on the use of carbon (coal), coke, and graphite. The manufacturing techniques of all three types are roughly the same, but a range of raw materials is used for producing the various qualities of graphitic and carbon refractories.

Anthracite coke is one important carbon refractory material and is obtained by heating anthracite coal at about 1 300°C for up to 24 h in shaft furnaces. Coke obtained by heating other coals without oxygen in coke ovens is also used for making refractories. Petroleum and pitch cokes are used; the most important for the refractories producer are those with low ash contents. Petroleum cokes are solid carbonaceous products obtained from petroleum residues by heating at 500–750°C.

The properties of three types of coke of interest to refractory producers are given in the following table:

|  | Petroleum Coke | Coal Coke | Crucible Graphite |
|---|---|---|---|
| Ash content | 0·3–0·75 | 10–13 | 8–12 |
| Sulphur | $<1\%$ | 0·5–1·3 | $<2$ |
| Volatiles | $<7\%$ | 1·6–1·6 | $<2$ |
| Moisture content | $<3\%$ | $<4$ | $<1$ |

Graphite is usually used in the crystalline or 'silver-graphite' form and the amorphous graphite form. Natural graphites and blast furnace graphite come under the first class, while the second class includes many other natural graphites as well as artificial graphite. Graphite is a very important refractory material because of its high resistance at elevated temperatures under reducing conditions.

Most carbon materials, apart from massive graphite, are used as refractories after being bonded with various types of pitch and tar, followed by heat processing. The characteristics of these bonds are discussed in Chapter 15.

Clay–graphite mixtures are commonly used for a variety of refractory applications and are made from plastic fireclay and graphite or other carbonaceous material.

## ZIRCON REFRACTORY RAW MATERIALS

Zirconium compounds suitable for making refractories include zirconia, $ZrO_2$, zirconium silicate, $ZrSiO_4$, and some less common compounds such as calcium and strontium zirconates.

Owing to the phase inversions taking place when zirconium compounds are heated, giving rise to volume changes during the firing and service of the refractories, zirconium materials normally have to be stabilised by combination with small amounts of such materials as calcium oxide, cerium oxide, yttrium oxide, etc. (*see* Chapter 14).

Zircon refractories are made both from granular zirconium compounds and also by fusion and subsequent casting into blocks. Various combinations of alumina–zircon, mullite–zircon and others are used as refractory materials, and these are of special significance in the glass industry. The most common raw mineral for obtaining zirconium compounds is baddeleyite.

## NON-OXIDE REFRACTORIES

Under this classification we may include such specialised materials as nitrides, silicides, and carbides. The raw materials employed in their fabrication are specially processed by the chemical industry. Many of the non-oxide refractories are very hard and have high melting points and are often used in cermets, that is, combinations of metals and ceramics. They are finding increasing use as engineering materials for making special compounds for use in gas turbines, rocketry and guided missile systems, as well as in the chemical industry where they find specialised uses for containers for corrosive materials.

*Chapter 4*

# How Refractories are Made

Ceramic and refractory materials are usually fabricated by grinding the raw materials, grading the resulting grains and then reassembling them by suitable bonding techniques into articles that possess predetermined properties. A firebrick, for instance, is made by blending variously sized grains of grog (chamotte) with plastic clay, followed by fabrication, drying and firing to render the bonded mass strong and resistant to furnace conditions.

The technical procedure, as outlined here, is basically simple but the various permutations of materials, grain-size distributions and bonding methods and also firing temperatures and subsequent treatment, make up the very complicated subject of the fabrication technology of refractories. Other methods of fabrication such as fusion casting and slip casting are discussed below. The various processes involved in making refractories are discussed under various headings.

## GRINDING AND GRADING MATERIALS

The raw materials delivered to the factory may be either non-plastic or plastic. Fired (calcined) grogs or alumina, for instance, are non-plastics, whereas natural clays are plastic. The lumps of raw materials need to be ground in various machines such as in ballmills or tubemills and the resulting powders graded by screening or similar methods, and then used in the batch for making the product (*see* Figs. 4.1–4.3).

When a material is crushed or milled the surface area is increased and this increases the surface energies of the resulting particles. The capacity of powdered materials for reaction during heating, either between themselves or with other materials in the batch, is an important aspect of refractory science. For instance, the surface properties of colloidal materials, some of which are present in many clays, are very important and determine the rheological properties of the systems containing them (the viscosity, fluidity, thixotropy, etc.). The blending properties, the

Fig. 4.1. Roll crushers are used in various sizes and designs for the primary crushing of fireclays. *Photo:* Courtesy William Johnson & Sons (Leeds) Ltd.

Fig. 4.2. Wet grinding rod mills such as this 10-ft long unit are used for milling grogs, calcined bauxites and other hard materials. *Photo:* Courtesy William Johnson & Sons (Leeds) Ltd.

mechanical strength, the behaviour during drying and firing and the final
physicochemical properties of the refractories depend very closely on the
surface area of the original components of the batch.

The grinding and milling of refractory raw materials is therefore a vital
stage in the production cycle. Much research continues to be done into
milling and various techniques are employed to accelerate the process
and bring the surfaces to a state suitable for subsequent firing reactions.

FIG. 4.3.    A 21-ft long compound ball tube mill used for milling calcined and other
hard materials in the refractories industry. *Photo:* Courtesy William Johnson & Sons
(Leeds) Ltd.

The grindability of materials depends not only on the hardness and
strength of the lumps but also on their proneness to cracking; the unifor-
mity of their particles; their plasticity and other properties. The economics
of milling, an important factor, depend also on the size of the lumps of
material; the type of mill employed; the ratio of grinding media (mill
pebbles or balls) to the charge and amount of water and other factors, such
as the peripheral speed of the mill.

Table 4.1 indicates the degrees of grindability of some common ceramic
raw materials. The quartz sand with a factor of 0·6–0·7 is the most difficult
to grind and the dry clay with a factor of 1·51–2·03 is the easiest.

These grinding factors can be used to determine the output of mills
handling different materials; for instance, if a ballmill is used to grind

TABLE 4.1

*Grinding Factors for Raw Materials*

| | |
|---|---|
| Quartz sand | 0·60–0·70 |
| Coal | 0·70–1·34 |
| Magnesite | 0·69–0·99 |
| Cement clinker | 0·80–1·00 |
| Limestone | 0·80–1·10 |
| Blast-furnace slag | 1·00 |
| Talc | 1·04–2·02 |
| Lime | 1·64 |
| Dry clay | 1·51–2·03 |

limestone and gives an output of 15 ton/h, then the same mill when grinding lime will give an output of

$$15 \frac{1·64}{1·0} = 24·6 \text{ ton/h}$$

where 1·64 and 1·00 are the grinding coefficients for lime and limestone respectively.

Hardness depressors are substances, *e.g.* certain silicones, added to the mill to reduce the hardness of the charge and hence accelerate grinding. The theory behind this action is rather complicated. Simply, what happens is that the liquid enters the numerous microcracks existing on all normal surfaces of solids, opens up these cracks and exerts very high pressures inside the particles. In this way the particles are broken down and grinding occurs much faster. Besides silicones which are of recent use, hardness reducers have been used for many years in the form of common salt, calcium, magnesium and aluminium chlorides and various surface-active agents such as tannin and petroleum waste. The proportions required are very small, normally 0·02–0·2% by weight.

Screening follows grinding. Its aim is to produce various classes of grain sizes which can subsequently be reconstituted to make a batch of the desired particle-size distribution. Some mills operate continuously and incorporate continuous screening and classification systems. The particle sizes of clays being processed may be checked and classified by air separation techniques.

The manner in which the particles of non-plastics and plastics are put together to make up a refractory product obviously has a critical effect on many properties, including density, porosity, and consequently slag resistance and gas permeability. It is not only the refractory (PCE and refractoriness-under-load) properties that determine the ultimate life of a product, but also the physical properties, including grain-size distribution and physical structure.

## BLENDING BATCHES

The purpose of blending the raw materials to make a refractory batch is to transform the solid components, which have different grain-size compositions, and the liquid additions (water, and solutions of such materials as sulphite lye or other cellulose derivations used as temporary, or green, bonds) into a macrohomogeneous mixture that can be subsequently moulded or shaped by one of the numerous fabrication methods employed by modern refractory manufacturers.

Blending of the particles of raw materials is only one of the processes in the continuous chain of processes, starting with the mining or winning of the raw material, and finishing up with the reconstituted fired mixture of these materials.

Blending depends on many properties and factors. Not many years ago it was the practice to age or sour the clays and other raw materials with the aim of evening out the properties of these materials, but in recent years the tempo of industrial production has increased so much that these time-consuming processes are now largely ignored; the aim is to deliver raw materials straight from the mines or quarries and involve them in continuous processing which terminates in the production of the final refractory material ready for furnace construction.

The speed and the final result of blending is largely determined by the particle shape and size; the overall grain-size composition and that of each component; the total number of materials being blended and their ratios; the frictional properties of the particles with respect to each other; the sticking capacity of the materials in the pans and the mills; the presence and amount of water and adhesive components such as sulphite lye and the final degree of 'grinding' the materials taking place in mixing. From this it is obvious that blending has a critical effect on the properties of refractory materials.

A wide range of mixing machinery is now available to refractory manufacturers and the choice of a particular model depends on the purpose in hand. The efficiency of the mill is governed by the design, especially the critical speed at which it can be operated; the type of movement of the particles and the variations in grain-size composition which take place during blending as a result of attrition and crushing, for instance under the rollers of runnermills.

The aim is to produce a homogeneous body. This is largely determined by the method of feeding the raw materials into the mixer. For instance, premixing will often give different results from those obtained when the components are fed individually. The precise points at which the water, clay slip, or solutions of cellulose derivatives (used as the temporary bond) are added will also have an important bearing on the final product.

Because of the nature of the raw materials delivered to the refractory production plants and in spite of the fact that the design of the mixers and certain other factors remain constant, the product often varies because of many other uncontrollable factors, or because previous processing stages cannot be accurately controlled. Blending is therefore often a matter of chance; precision in predicting final results is not high.

In the past it was widely thought that the blending of, for instance, fireclays, grogs, sulphite lye and other materials used in refractory batches was simply a physical process. However, it is now known that important chemical surface reactions occur between the colloidal constituents of the raw (green) batch materials and the non-plastics. The precise manner in which colloidal particles adhere to, and form films around, the particles of calcined fireclay and other non-plastics in the mix, has an important effect on subsequent heat processing.

The production of a refractory batch ready for moulding and firing can be said to comprise two basic inter-related stages:

1. The blending of the components and mill additions, and
2. The physicochemical reactions, which largely occur in the colloidal constituents, and which will help in subsequent fabrication by pressing, slip casting, vibration moulding, hot pressing, or the particular method to be employed.

The refractories manufacturer pays close attention to blending of batches. It is not yet known how water, one of the raw materials in the ceramics industry, reacts with such materials as silicon carbide, chromites, and other materials. Some research has been done on reactions that occur between aluminium oxide and electrofused materials.

It is too often thought that water is an inert medium which is useful merely as a modifier of the physical form of refractory batches. Research is now indicating that water is a vital reactant with batch materials and must be considered from its physicochemical aspect as well as its purely physical aspect in modifying the plasticity and other rheological properties of the batches.

## FABRICATION TECHNIQUES

The aim of moulding refractory batches is to produce a brick, block or special shape that can be handled, placed for firing and subsequently heat processed to yield a permanent, fully stable refractory suitable for furnace construction and other heat-resisting purposes. The batches used for moulding may be fully plastic, semi-plastic, powdered, or liquid (ceramic slips). Low-plasticity or non-plastic bodies normally contain 3–7% moisture and are called semi-dry bodies (in some countries 'dry-press' bodies). Plastic bodies contain 16–22% moisture and casting slips up to 45% moisture (Fig. 4.4).

FIG. 4.4.   Belt-type apron feeders are used as ancillary machines with presses for carrying ground and prepared materials from mixers to the presses. *Photo:* William Johnson & Sons (Leeds) Ltd.

Other methods now being used to fabricate refractories include iso-static (hydrostatic) pressing; vibration moulding; hot pressing and the casting of hot plasticised slips. The casting of hot melts such as combinations of alumina, mullite, and zirconia is also of growing significance for producing solid homogeneous blocks of highly refractory materials that are of special interest to the glass industry. These methods will now be briefly described.

Dry pressing (semi-dry pressing) is used with a range of press designs (*see* Figs. 4.5–4.8) for making standard and complicated shapes for many purposes. The mould is filled with a weighed quantity of press powder which has been carefully blended and then pressed and compacted with a plunger lowered from above on to the material. Double-sided pressing is also practised; two plungers are used, one from each direction, vertically upward and downward. The full pressing force may be applied at once in a single stroke, or in steps. When the pressure is released the pressed article is ejected from the mould and is then sent for subsequent processing. The compaction or densification of the particles of materials in the batch depends on the grain-size composition; the rheological properties of the bond and the system as a whole; the pressing force, and whether the pressing is double-sided, and staged or simple.

FIG. 4.5. Hydraulic presses designate RPB have pressing forces ranging from 10 to 20 tons and are used for shaping stoppers and nozzles from extruded slugs for the metallurgical industry. *Photo:* Courtesy Bradley and Craven Ltd.

In designing press moulds the engineer must take into account that the final fired sizes of the product will be different from those of the green unfired product. Immediately upon release from the mould some pressings tend to expand slightly owing to the elastic expansion of the materials. Subsequently, upon drying, the brick or shape will shrink. The moulds

FIG. 4.6.   A 600-ton mechanical toggle press such as this can be used for shaping many dry-press bodies such as silica, sillimanite, fireclay. *Photo:* Courtesy William Johnson & Sons (Leeds) Ltd.

must be designed on the basis of empirical data since the raw materials are so variable that no strict theory can be universally applied to all forms of refractory material.

The two main types of presses used for dry pressing in the refractories industry are hydraulic presses and friction presses.

Plastic or wet moulding is normally used for making complicated shapes from plastic bodies containing moderate or large quantities of clay. The mixture containing at least 16–17% moisture is passed through a pugmill fitted with vacuum devices to extract the air and is delivered for subsequent forming in a continuous slug of the required cross section. Wire cutters or similar devices are used to slice sections from the mouth of the pug. These blanks are then finished off on finishing presses, which give them their final shape and accurate dimensions.

FIG. 4.7. Automatic sleeve press. *Photo:* Courtesy Bradley and Craven Ltd.

The method is very similar to that used in the production of building brick except that the second stage of finish-moulding is left out in the production of common brick. The sliced slugs are delivered to the drier, followed by firing. Common firebrick is also treated in the same way as common building brick. The production rates for the simple pugging, cutting and drying-firing cycle are very high and the process is easily automated.

The finish-pressing is applied to higher quality refractories because they need to be accurately shaped and sized and have smooth surfaces to help in building up a regular, monolithic structure in the furnace lining. The pressing forces applied in moulding by this technique are from 10 to 40 kg/cm$^2$. Naturally the structure and hence the properties of the extruded (straight from the pug) product and of the product subsequently pressed in a mould, will be different; care is taken in the factory to retain the structure of the original pugging.

FIG. 4.8.    500-ton hydraulic press. *Photo:* Courtesy Bradley and Craven Ltd.

**Isostatic Pressing (Hydrostatic Pressing)**
Although this method of shaping ceramics was developed almost 60 years ago it has not found wide use in the refractories industry. Research is being done into its use for shaping sanitaryware and electrical porcelain. If an economically commercial process could be developed it could offer important advantages over existing techniques for the refractories industry.

The process is basically simple and consists of placing powder in a thin rubber or plastic shell (the mould) and then evacuating this shell so as to fabricate the product. The shell has the shape of the desired final article. The moisture content of the powder used for isostatic moulding is about 5%. One of the main advantages of the method is that the density throughout the pressed item is uniform and there is no need for subsequent machining of the pressing.

The shell containing the powder or precompressed blank is placed in a water emulsion, oil, glycerine, or similar liquid and pressed by forcing out the liquid or applying a load to a piston. The pressure applied in this way varies very widely and various techniques involve pressures from 250 to 700 kg/cm$^2$.

Hot, warm, and cold hydrostatic pressing techniques have been described in the literature, the temperature ranges being from room temperature to 80–100°C and as high as 1 000–1 600°C.

More information about the isostatic pressing method for refractory shaping can be found in the References which cover American, German and Soviet researches in the subject.

**Vibration moulding**

A common aim in pressing and moulding refractory batches is to maximise the density in the green state so that upon firing the article becomes a homogeneous, high-density product, resistant to the action of molten metals, slags and glasses. This process can often be helped by vibrating the particles of the batch during fabrication. The particles are moistened, poured into a mould and then vibrated and pressed. Vibration gives the particles a better chance of taking up positions relative to each other that are more conducive to the formation of dense structures.

During pressing densification is achieved by friction and mutual bonding. However, if vibration is applied, the particles can move much more freely and will therefore assume a denser structure.

This technique was originally developed for fabricating powder metals and it is in this technology that it has found the greatest use. However, non-metal refractory technologists are also interested in it, and the principles involved will now be briefly described.

The fabrication techniques now being used by powder metallurgists include rolling; slip casting; die pressing and extrusion; isostatic pressing; hot pressing and impregnating porous refractory frameworks with fusible metals; explosive and vacuum fabrication methods; pressing metal fibres, and centrifugal shaping techniques. Not one of these techniques is free from all the shortcomings limiting the use of powder-metal methods. Furthermore, many of them are extremely complicated and do not find ready use commercially.

It is important that the refractories producer possesses a method for shaping very complicated products with uniform densities and other properties. This is where vibration compaction may be of value.

The problem of producing high-duty construction materials, particularly those possessing high densities and strengths, from any powdered materials of any dimension and shape, can be solved if it is possible to obtain a sufficiently uniform, unstressed, strong packing of the particles with the required density (Shatalov *et al.*, 1965). The method of tackling this problem is based on the principles of physicochemical mechanics, which cover methods of controlling the structure formation in the production of porous and compact materials made from a great variety of substances.

One of the principles of physicochemical mechanics is that of limiting the rupture of the original bonds between the structural elements in order to achieve the desired uniform distribution and dense packing. The idea is being used in the blending and compaction of powdered materials and concentrated disperse mixtures with the minimum quantity of bond. Examples include concrete and ceramic mixtures which are processed under vibration. Vibration contributes to better orientation, breaks up friable 'bridge' structures and hence ensures the packing of the particles at moderate pressures.

This is the foundation of vibration moulding which may find extensive use in the fabrication of refractory materials.

When the powdered batch is poured into the mould, vibrated and pressed at the optimum pressing force, the product is found to have very little or no elastic expansion (backspring) and this is one distinct advantage over other methods of fabricating refractories. Most of the information about vibration moulding is to be found in the literature on powder metals and so far very little work has been done on other refractory materials.

Vibrations may be applied to the batch by a press plunger or through the mould itself. If the plunger is used the layer of powder adjacent to it is rapidly densified and this is quite satisfactory when the product is not very thick. However, when thick articles are being moulded the oscillation rate is retarded and the density of the particle packing declines from the top to the bottom of the brick. It follows that the mould and press must be selected to match the size and shape of the product.

As in other moulding methods used for refractories, the water content is important. In the shaping of metal powders it is found that the optimum moisture content is that corresponding to the water absorption of the pressed article. Any excess will tend to reduce the density of the product and retard the production cycle.

The vibration moulding technique is subject to much more accurate control than other techniques. However, segregation of the aggregate particles and bond materials, or of particles of different materials, may occur if the optimum cycle is not maintained. For instance, very high amplitudes; the presence of certain lubricating and surface-active agents in the batch; too high a moisture content and other factors may be detrimental to the final product. Vibration-moulded brick may have a very high strength; low air shrinkage; a high degree of isotropy and a very uniform structure, providing the technique is properly applied.

The advantages of this shaping method for the user of refractories are that he buys a much superior product, with closer tolerances and a better working life. A very wide range of materials can be so shaped, including oxides, carbides, nitrides, and also high-grog aluminosilicate bodies.

For the manufacturer the advantages, in addition to those mentioned above, include a much faster production cycle, since the moulded goods can be placed directly on the kiln cars and fired to give precision shapes that need no subsequent machining. Very complicated articles, including hollow products and tubes, can be shaped by this technique. Vibration techniques may also be useful for granulating, blending, and screening powders, and also for moving powdered materials from one working site to another.

### Slip casting

The shaping of ceramics by using water and other liquid suspensions of blends of non-plastics and plastic clays is a very old method, and is widely used for fabricating tableware, sanitaryware, and other products (Fig.

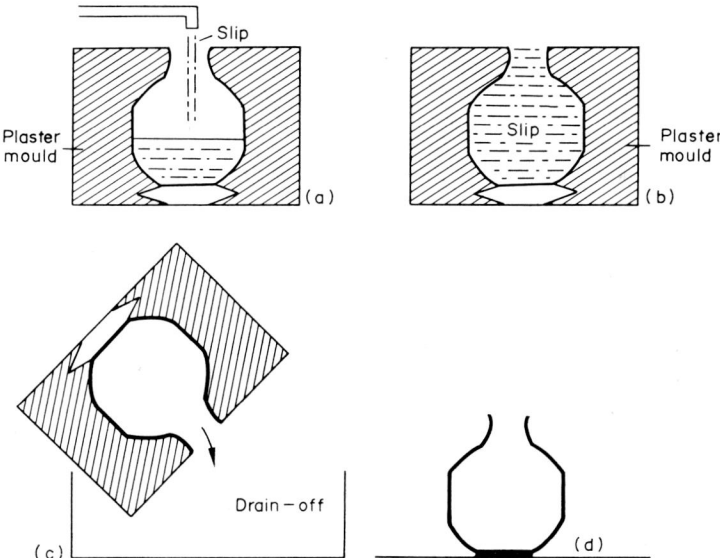

FIG. 4.9. Slip casting hollow ware (crucibles, tubes, etc.). (a) Plaster mould is filled with slip; (b) the absorbent plaster sucks water from the slip, building up the cast (casting-up process); (c) excess slip is drained off; (d) finished cast removed from the mould.

4.9). It is used in the refractories industry for making certain complicated shapes but it has the disadvantage of producing products with density variations owing to the segregation of the solids during settling.

The method consists of grinding the raw materials in water or other medium, adding certain deflocculating agents so that high-density slips can be used, and casting the resulting slurries or slips into moulds made of plaster of paris or other materials, such as easy-fired clay. The body is allowed to cast up, and then the article is taken from the mould, dried and fired.

In the refractories industry the method is used for making thin-walled products and complicated shapes. It is used for making oxide refractories, magnesia-spinel products, silicon carbide, borides, silicides, and other high-duty materials.

The casting slip itself is made up of finely milled refractory particles and consists of a microheterogeneous system comprising a finely dispersed solid phase and a liquid phase. The properties of the slip are very similar to those of colloidal systems. Casting slips are not Newtonian liquids and therefore their rheology is somewhat more complicated than that of such liquids. The fluidity, viscosity and thixotropic properties must be considered with special reference to ceramic technology.

Clay slips are sometimes used as enveloping films for high-grog mixtures. In this case a slip is made of plastic clay and this is blended with the non-plastics in the form of calcined clay, alumina, aluminosilicate or other materials, the resulting batches subsequently being fabricated by pressing and then dried and fired. The clay in the slip is used as the bond. This sticks the particles of non-plastic together during firing. By using clay slips it is possible to distribute the clay bond much more uniformly over the particles of grog than if the clay is added dry to the batch.

The simple process by which a clay slip or suspension of non-plastics stabilised with a colloidal system is cast up into a solid article is illustrated in Fig. 4.2.

### Casting oxides by the slip method

Most oxides of interest to refractories technologists can be cast in the form of water suspensions since they do not react with water. Calcium oxide, however, does react and it is necessary to use non-water media. Alumina and zirconia are commonly fabricated by casting and most of the information published on the slip casting of oxides refers to these two materials.

Crucibles, tubes, thermocouple sheaths and other special components are made by this method. Crucible mixtures may consist of alumina with small additions of titanium dioxide or a mixture of $TiO_2$ (1%) and about 2% $ZrO_2$. This improves sintering and increases the thermal-shock resistance of the crucibles. These additives do not affect the rheological properties of the slip.

The normal method is to prefire the alumina, grind it finely, remove the iron, and then vacuum-treat the slip before casting it in plaster moulds. The products removed from the moulds are fettled, dried and then fired. The alumina is calcined at 1 450–1 600°C, the precise temperature depending on the crystal sizes of the oxide. It is ground in vibromills or ballmills by dry or wet methods and provision has to be made so that the materials do not become contaminated by the mill linings. Rubber, alumina, or porcelain linings are therefore used.

Because of the low mechanical strength of unfired alumina and zirconia products, organics such as polyvinyl alcohol and certain resins are frequently added to the slips. This increases the green strength. The green products are placed in saggars that are packed with alumina or zirconium dioxide and fired at 1 680–1 710°C, holding for 2–10 h, depending on the sizes and the required final density. Products containing $TiO_2$ additions are fired at 1 550–1 600°C. Well-sintered alumina crucibles are similar in appearance to porcelain, and defects are easily seen by visual examination.

Zirconia crucibles and similar products may be used at temperatures of up to 2 500°C. Owing to the polymorphism of zirconia (it exists in two forms: monoclinic and tetragonal) it needs to be stabilised and this is done by making additions of 4–6% by weight of magnesia or calcium oxide, or in some cases ceria or yttria, which form a continuous series of solid solutions, crystallising in the cubic form.

The casting technique for zirconia is very similar to that for alumina. More information on the properties and uses of cast alumina zirconia, beryllia, ceria, and other oxide refractories will be found in Chapter 16.

**Hot pressing**

The refractories technologist employs hot pressing to simultaneously densify and strengthen powders as a result of pressure and sintering. The technique is widely used in making cermets, in the powder metals industry and for fabricating oxide refractories. Its main advantages over other moulding methods include the appreciable reduction in the consolidation time (from several hours to minutes) and the use of very active and fine starting materials without any preprocessing. The resulting products possess a density very close to the theoretical density of the materials; this is usually unobtainable by ordinary sintering of most materials.

The materials which can be fabricated by hot pressing include silicon and boron carbides, boron nitride and other highly refractory compounds. More recently cermets, oxides and other types of ceramics have been so fabricated. Because of the special equipment required the technique can only be used for high-duty, and high-cost products. It is usually restricted to those materials that cannot be produced with near-theoretical densities by other methods.

The chief advantage of hot pressing from the manufacturer's point of

view is that it can be done at temperatures close to the sintering temperature in the free unloaded state. The chief stage of the reaction inside the
mould during hot pressing is recrystallisation; grain growth of the material
being sintered does not even commence. As a result the sintered article has
a very fine grain structure and a high relative density. Only simple shapes
can be treated by this method and these are then often machined into
special shapes.

The equipment comprises a graphite mould and a graphite plunger
housed in an induction furnace. The graphite mould itself is sometimes
used as the heating element. The pressing temperature varies from 200 to
1 200°C, depending on the material being pressed.

If graphite moulds are used the fabrication pressure cannot exceed about
250 kg/cm$^2$, so the process is carried out at relatively low temperatures
and moderate pressures. Alumina moulds have been developed, in which
case the pressures may rise to 700–1 400 kg/cm$^2$ to yield products with
the theoretical density at 1 100°C or even lower. Very high pressures of
1 000–1 100 kg/cm$^2$ and very high temperatures have also been used
(Alligro *et al.*, 1956).

**Fusion casting**
This process is very similar to the melting and pouring of glass. The raw
materials are melted, homogenised, and then cast into moulds. The resulting cooled products are refractories of very high quality and cost, suitable
for building furnaces such as those used for melting glass, stone castings,
slags, etc. The raw materials employed are alumina, aluminosilicate–
zirconium oxide mixtures, bauxites, diaspores, zircon and materials of the
sillimanite group. Various types of furnaces are used, the commonest
being three-phase tilting arc furnaces. The main problem in casting the
molten refractory materials is to obtain blocks of the required size and
shape, with a very dense structure, uniform and homogeneous, without
zoning or coarse texture. The casts should be as accurately sized as possible
to reduce machining since the material is very hard and abrasive and not
readily cut.

The casting of aluminosilicate fusions is more difficult than the casting
of most metals since it must be done at higher temperatures and when the
melt is poured from the furnace rapid cooling and hardening of the fusion
may take place, causing a variety of faults in the product. Furthermore,
molten refractories are much more viscous and the shrinkage much higher
than that of metals.

The materials from which the moulds are made are obviously very
important in this process. Clay-bonded quartz sand is often used; the clay
may be replaced by sulphite lye, various oils, or special bonds.

More details of the properties and applications of fusion cast refractories
will be found in Chapter 14.

## Casting thermoplasticised slips

In contrast to casting in plaster moulds using water suspensions at room temperature, this technique employs substances that are solid at room temperature and liquid at moderately high temperatures. Thus, by applying heat to the mixtures of solid and 'liquid' phases they can be made to flow and assume the shape of the moulds. When the article has been so fabricated, the mould is allowed to cool, whereupon the item solidifies and can be taken out for subsequent processing.

The technique is quite simple and much faster than casting with water slips since there is no need to remove the liquid phase from the mould or article, and solidified product can be obtained very soon after the casting process is completed. The green bond (wax or other organics) is burnt out when the article is fired.

The advantages of the method, in addition to speed, are that very complicated articles which can be machined after fabrication can be made; the use of accurately fabricated durable metal moulds; and the possibility of mechanising and even automating the casting operations. Disadvantages include the relatively high porosity of the casting because of the need to remove the organic material in firing; the expense of removing this plasticiser (normally into an adsorbent for subsequent regeneration) and the high shrinkage following heat treatment of the products.

Various modifications of this technique are in use, the commonest being plasticised casting under pressure. This reduces the amount of plasticiser to a minimum, and accurately sized and shaped products are obtained.

The common organics employed are paraffins and compounds such as oleic acids which improve the rheological properties of the system. An obvious essential property of the mixture is that the solid particles (the refractory constituent) of the mix must be readily wetted by the molten wax during forming, otherwise the product will not be uniform.

## FIRING AND SINTERING

The meanings of the words 'ceramics' and 'refractory' signify heat, resistance to heat, and the behaviour of materials when heated. In the production of refractory materials the final stage is normally firing or sintering at high temperatures. The working response of the material to temperature (and other furnace conditions) is ultimately determined by the essential character of the refractory material itself. The manner in which refractories are fired from their green, raw material state involves not only temperature but also time and furnace atmosphere.

In the early days of the refractory materials industry, firing was an empirical process in which the materials were placed in a kiln, subjected to heat and an attempt made on the part of the kiln operator to turn out hard-fired, dense products, free from cracks and other obvious defects. Today the process is far more complicated and the approach of the firing operator far more sophisticated.

From the simple 'burning' or firing processes, the refractories manufacturer has now advanced to the stage where his vocabulary includes such terms as oxidising fire, reducing fire, neutral fire, sintering, liquid sintering, solid-phase sintering, and many other specialised terms.

The firing process involves a multitude of complicated, overlapping physicochemical processes. The changes taking place in, for instance, a firebrick or mullite block, a magnesite–chromite shape or a zircon nozzle during firing include (in addition to obviously physical changes such as thermal expansion, the change of solids into liquids, and temperature changes) much less obvious processes such as solid reactions, solid solutions and phase transformations, all of which determine the ultimate success of the application of heat in producing a satisfactory refractory material.

All ceramics have been subjected to some form of heat treatment. However, the refractory ceramic will be subjected to heating during its useful life. A piece of tableware, for example, or a wine glass, has been subjected to the ordeal by fire, and owes its properties to this treatment. In addition to being fashioned by fire, refractory materials must also live by fire. The manner in which the refractory material has behaved during its original firing often determines the way in which it behaves during service in a kiln or furnace lining.

It is true, of course, that some refractory materials are not fired before use. These products, known as unfired refractories, may take the form of shaped bricks or blocks, castables, concretes and guncreting mixtures which are placed in position in the plastic or semi-plastic state, and undergo their 'fabrication' firing at the same time as, or just prior to, the service firing.

## Conditions for firing

The firing temperatures for refractories will be decided by the nature of the raw materials, the ultimate purpose of the product, and other conditions.

The firing atmosphere will be determined by the nature of the raw materials and the need for developing certain types of compound in the product being fired. Sometimes the atmosphere cannot be strictly controlled and this will have a serious effect on the finished product. The use of purer forms of fuel and especially the elimination of sulphur from gas and oil and the declining use of high sulphur coals, has improved the atmospheric control inside furnaces and kilns burning refractories.

The firing time is determined by the nature of the material, and the final desired state of the refractory product (including its density and porosity, and the precise stage of the so-called 'arrested reaction') that must be reached before the product is considered to be fired.

An important factor here is the cost of firing. Any method of preparing a raw material that can be developed into its 'matured refractory' state at lower temperatures is of economical benefit to the manufacturer and user. To quote a simple example, fine grinding of most refractory raw materials causes firing reactions to take place much more quickly than when the raw materials are coarsely ground. Thus, in fabricating most refractories the manufacturer aims to load them into the kiln in the most physically suitable state for the completion of the essential firing reactions.

The features of the heat-treating of various types of refractories such as silica brick, fireclay, chrome-magnesite, pure alumina, fusion cast, etc. are numerous and varied. In order to indicate simply and briefly the processes taking place in firing, brief accounts will now be given of some of the processes that take place in many of the materials well known to the refractories user.

**Firing principles**
When the dried product is loaded on to, say, a tunnel kiln car, or placed in a periodic kiln, and heated, numerous physicochemical processes occur, including oxidation of the organic impurities such as coal; decomposition of carbonates, sulphates and other salts; dehydration, which involves the removal of the physically bonded water and also the water of crystallisation in the clay; reactions in the solid phase; phase changes and the production of new phases from the breakdown products; the development of solid solutions and melts; recrystallisation sintering; increases in strength, and several other processes.

The mechanical strength of the product being fired may vary, and at certain critical temperatures special care must be taken in case the expulsion of steam and other gases causes the product to rupture.

The firing process is not simply a matter of applying a steady supply of heat and then cooling. Different stages of the process require different quantities of heat input, since the reactions occurring require different quantities of heat. In some cases reactions actually generate heat (exothermic reactions) outside that generated by the burning of fuel in the kiln.

Cooling is also another critical stage in the process. In fact, several important processes, including crystal inversions, crystallisation and the formation of cracks and microcracks in the products, occur during cooling. For example, in the production of fusion cast refractories, the cooling process may determine the success or otherwise of the whole process. The fusion-cast blocks are normally annealed in the same way as glass articles are annealed to remove any damaging stresses developed during cooling.

The behaviour of ceramic materials during heating constitutes a major section of the subject of ceramics and refractories science. The user of refractories is mainly concerned with the behaviour of the product when it has been converted from raw material into ceramic material. However, some understanding of the behaviour of raw materials in firing is essential for the complete understanding of refractory materials in service. Ceramic texts may be consulted for explanations of this behaviour and special attention should be paid to the firing behaviour of fireclays, kaolins, magnesites, alumina, aluminosilicates, and the other materials mentioned in Chapter 3 of this book. Some suitable books are listed at the end of this chapter.

**Unfired refractories**

As mentioned above, kiln and furnace builders use unfired refractories in addition to fired bricks and shapes. These include concretes, castables and guncreting mixtures which can be applied to the furnace *in situ* (*see* Chapter 18). Guncretes are refractory concretes of varying composition formulated for specific purposes in the plant of the refractories manu-facturer, or on site and delivered to the user ready for use, or as powders that simply need the addition of water before application. The main constituents are grogs or other refractory aggregates, refractory (usually high-alumina) cements, and possibly organic bonds such as sulphite lye and resins.

The aggregate or grog may be identical with the materials used for making fired refractories, or may be specially made for concretes and castables. For example, a material known as 'bubble' alumina is widely used in insulation. This is a high-$Al_2O_3$ material specially processed to give low-density fired products due to the air entrapped in the bubbles of alumina. It is used as hot-face insulation.

The working properties of refractory products are very similar to those of building concretes, except, of course, that the refractory materials have a much higher fusion point (normally above 1 600°C). The advantages of unfired products include their ability to form monolithic linings without joints. When precast refractory blocks are used, the number of joints is considerably reduced. The products are usually fired in an oxidising atmosphere in the furnace where they are to be used, but subsequently reducing atmospheres may prevail, as in the use of ordinary fired refractories.

The types of refractory concretes include those based on Portland cement, alumina and high-alumina cements (of the Secar and Lafarge types), water glass for bonding purposes and periclase cement (magnesite finely ground and slaked with aqueous solutions of magnesium chloride and hydrated magnesium sulphate). Phosphate bonds, dinas concretes with various types of bond, aluminosilicate concretes, and basic concretes are also in use. In other words, all the materials used in fired refractory products can be employed with varying degrees of success in the formula-lation of refractory concretes and castables. The properties and applica-tions of concretes and castables are described in Chapters 18 and 24.

# REFERENCES

Alligro, R. E. *et al.* (1956). *J. Amer. Ceram. Soc.* **39**, No. 11, 387.

Bal'shin M. Yu. *et al.* (1961). *Doklady Academy of Sciences, USSR* **136**, No. 2, 332.

Blokh, G. S. *et al.* (1955). *Steklo i Keramika* No. 6, 17.

Lohrengel, H. (1967). *Ber. deuts. keram. Ges* **44**, No. 3, 90.

McCreight, L. R. (1951). *American Ceramic Society Bulletin* No. 4, **127**.

Shatalov, I. G. *et al.* (1965). 'Physicochemical Principles of the Vibration Densification of Powders'. Nauka, Moscow, p. 4.

Sheinin, E. I. *et al.* (1965). *Steklo i Keramika* No. 8, 23.

Svec, J. (1964). *Ceramic Industry* **83**, No. 5, 55.

Wagner, H. E. *et al.* (1951). *American Ceramic Society Bulletin* No. 10, 341.

*Chapter 5*

# Classification of Refractories

The British standard specification of terms relating to refractory materials (BS 3446:1962) lists 52 types of refractory material but attempts no general classification. The problem of classifying refractories is very difficult since the factors to be taken into account are numerous and not necessarily related. For example, the materials may be classified in terms of service temperature; chemical composition; types of raw materials employed (raw, natural or synthetic) and physical properties such as porosity, shape and size.

## CHEMICAL COMPOSITION CLASSIFICATION

All known refractory materials may be classified into eight general groups based on chemical composition, as follows:

1. Silica (dinas and quartz)
2. Aluminosilicates (fireclay, siliceous, aluminous)
3. Magnesia (magnesite, forsterite, dolomite and spinel)
4. Chromite (chrome, chromite itself, chrome–magnesite and magnesite–chromite)
5. Zirconium compounds (zirconia and zirconium silicate)
6. Carbon (graphitic and coke-based)
7. Carbide and nitride (silicon carbide and others)
8. Oxide (pure-oxide such as alumina, ceria, beryllia, etc.).

Each of these chemical-composition classifications can of course be subdivided in various ways. For example, under aluminosilicate we may divide the materials into dense and highly porous (insulating brick). The basic refractories classified under magnesite dolomite, etc. may be either ordinary or tar-bonded. Zircon refractories may be fusion-cast or ceramic bonded.

## RAW AND SYNTHETIC CLASSIFICATION

The raw materials for making refractories can be divided into two groups:

1. Those found naturally and subject to certain processing at the quarry or in processing plant prior to use.
2. Synthetic materials such as mullite made from bauxitic rock and other minerals.

'Tonnage' refractories such as fireclay, magnesite and chrome–magnesite are all generally made from the natural raw materials, while the special, normally high cost refractories such as alumina, mullite, beryllia, the carbides, nitrides and silicides, etc. are generally fabricated from synthetic or man-made compounds produced by the chemical processing industry.

It is customary to classify refractories in terms of their PCE values, that is, the degree of refractoriness. For example, the materials fusing at temperatures between 1 580 and 1 770°C may be called simply refractory materials, those fusing between 1 770 and 2 000°C are called highly refractory, and the term super-refractory or high-duty refractory is normally reserved for materials fusing above 2 000°C.

Further classifications may be involved in describing refractories with reference to the methods used to fabricate them, for instance, pressed, vibrotamped, slip-cast or fusion-cast, etc. The materials may be fired or unfired (e.g. conventional firebrick, compared with guncrete mixtures and concretes).

Another classification term employed is that referring to the particular type of furnace or plant in which the refractory may find common use, for instance, steel-plant refractories, casting-pit refractories, electric-arc melting refractories, continuous casting-plant refractories, cement-kiln refractories, glass-tank refractories, etc.

The refractories producer and the user of refractories are continually investigating the advantages and properties of new materials and those developed by the industry for use in furnaces outside their own particular field. The cement technologist for instance is keenly interested in the use of basic refractories originally developed for the steel industry. The pottery kiln designer has in the recent past been very interested in finding a suitable material to replace silica brick in the roofs of pottery kilns.

Such interchange of ideas and research knowledge results in the continuous development of new and improved types of refractories for all industries. New combinations of raw materials, synthetic products, and intermediate materials, combined with new processing and firing methods, all lead to the development, testing and use of novel materials, and a gradual improvement in the working life of refractories.

The user of refractories is normally concerned with two conflicting aims. Firstly, to make the furnace linings last as long as possible and secondly to accelerate the processing of the furnace charge. The latter is usually achieved by raising the temperatures in the furnace and this, of course, means reducing the life of the refractories.

Not many years ago the policy in many heat using industries was to build furnace structures with large quantities of relatively cheap firebrick and other common refractories, work the furnaces as fast as was consistent with safety, demolish the structures and rebuild with fresh loads of cheap refractories. Today, rising labour costs have meant that the kiln and furnace builder must look for better refractories (and pay higher prices) which will either last much longer, once they have been built into the structure, or enable the user to work the furnaces much faster and at much higher temperatures. The emphasis in refractories manufacture and use can now be said to be on quality rather than quantity. This is illustrated by what has happened in the use of silica brick for the manufacture of gas. New gas-making techniques have meant that the silica brick manufacturer can no longer rely on selling massive quantities of silica brick to the gas industry, which now requires refractories of much higher quality for new gas-making techniques.

In other industries, for example the steel industry, the development of new metal producing techniques will emphasise this trend. For instance, spray steel making would involve the consumption of even fewer refractories. Again, in the steel industry it is predicted that the open hearth is becoming obsolete and the economics of basic oxygen furnaces are such that it might be more economical to replace the modern open hearth by these oxygen units. The continuous steel casting process which is gaining ground in America, Russia and Britain is another challenge to the bulk refractories producer since this streamlined technique does away with soaking pits and other intermediate processes. On the other hand, the continuous method requires superduty refractories such as zirconia and long-life tondish nozzles, etc.

From these brief remarks about refractory trends it will be seen that any new and improved process for producing metal fundamentally affects the type and quantity of refractories to be produced. The trend (expressed simply, at the risk of being imprecise) is away from the natural, unprocessed raw materials such as fireclay, quartz and sands, towards processed, synthetic and purified materials, particularly the oxides and synthetic mullite.

Special refractories such as carbides and silicides will also find bigger markets and more widespread uses for furnace building. Another marked trend is the greater use of monolithic linings built with concretes and other castables. This follows from the rising cost of furnace-building labour and the lack of the necessary skills to do this work.

In the following chapters (Chapters 6–18) details are given of the properties and distinctive production and application features of the various types of refractories. Section III will deal with the application of these refractories in various industries (Chapters 19–24).

*SECTION II*

# TYPES OF REFRACTORIES

## Chapter 6

# Fireclay–Chamotte Refractories

Fireclay refractories, such as firebricks, chamotte materials, siliceous fireclays, flint clays, and aluminous clay refractories, consist of aluminosilicates together with various amounts of silica making a total $SiO_2$ content of less than 78% and containing less than 38% $Al_2O_3$.

The term 'chamotte' needs a word of explanation. Originally a German expression, this term has now been adopted in the USA, Britain and elsewhere to denote a refractory clay that has been specially calcined and crushed for use as a non-plastic constituent of a refractory batch (Figs. 6.1–6.2). It is sometimes used as a synonym for the term 'grog'. The distinction, strictly speaking, is that grog is reserved to describe used or spent lining materials that have been recovered and crushed. In the German and Russian literature the terms *chamotte* and *shamot* respectively may mean firebrick.

### PHYSICOCHEMICAL PRINCIPLES OF FIRECLAY TECHNOLOGY

The manufacturing technique for fireclay (chamotte), kaolin and also semi-acid refractories, all of which are based on clays, is generally governed by the properties of the raw clays and the changes that occur during drying and firing when these products are shaped into firebrick. Another important factor is the ratio of the amounts of raw clay (the plastic constituent) and the non-plastic aggregate (the chamotte, or grog). Chamotte, as mentioned above, is in this case clay that has been calcined to render it relatively inert during firing. These materials are then blended to obtain the necessary grain-size distribution, ratios of the constituents, and the solids–water ratio in a state suitable for fabrication by the selected method, such as semi-dry pressing, wet moulding or slip casting, to yield a stable, chemically resistant and heat-resistant product known as a firebrick.

Naturally the final properties of the brick will depend on the chemical

Fig. 6.1.   A rotary chamotte-calcining kiln used to produce high-quality aggregate from plastic ballclays in Devon, England. The unit is controlled by the closed-circuit television system illustrated in Fig. 6.2. *Photo:* Courtesy Watts, Blake, Bearne & Co. Ltd.

and other properties of the raw materials, their ratios when blended, and also on the grain-size distribution and other factors. A very wide range of firebrick is commercially available. Normally the quality is considered to improve with increasing alumina content (but *see* below).

The physical properties, for instance the porosity, of firebrick, are very important. For example, if firebrick is used for lining steel-casting ladles it must be dense and resist the penetration of molten metal and slag. If, on the other hand, china clay is used to make a firebrick for insulating purposes, it is usual to render the product highly porous by adding

Fig. 6.2. Closed-circuit television is used to control chamotte production inside a rotary kiln. This prevents 'rings' forming in the kiln (*see* Chapter 20). *Photo:* Courtesy Watts, Blake, Bearne & Co. Ltd.

combustible substances such as sawdust, or by using air-entraining and foaming techniques in order to increase the air content of the product. By reducing the bulk density and increasing the apparent porosity of the product a better heat insulator is made than the dense firebrick selected for ladle linings.

The concentration and particle-size of grog or chamotte govern not only the shrinkage but also the mechanical strength, the spalling resistance, and the porosity.

The amount of water present during fabrication can also have a critical effect on the final properties of the brick. The chemical nature of the chamotte will of course be critical. For instance, if quartzitic materials such as gannister are used as the filler so that the brick expands during firing, then the shrinkage of fireclays can be compensated for and the final product will show an expansion after cooling.

## ALUMINA–SILICA PHASE DIAGRAM

The formulation, firing and service of fireclay–chamotte refractories is based on a knowledge of the alumina–silica phase diagram (*see* Fig. 6.3),

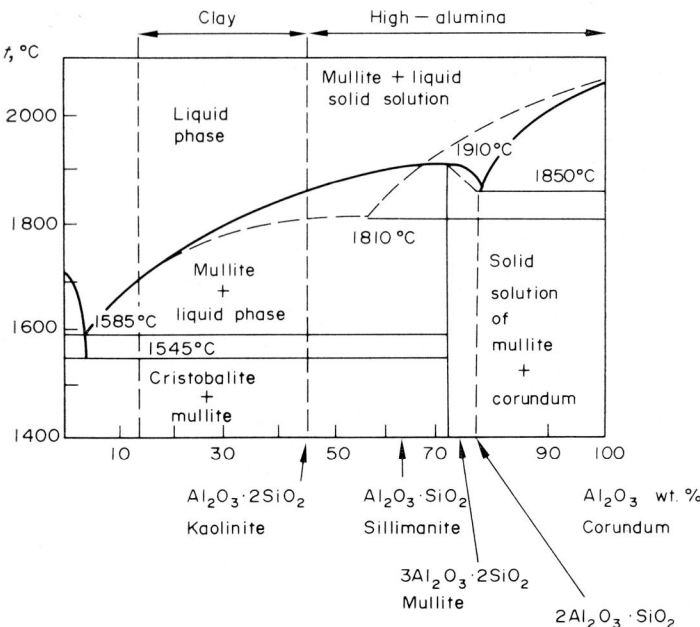

FIG. 6.3. Phase diagram for the $Al_2O_3$–$SiO_2$ system (broken lines according to Bowen and Greig; solid lines according to Toropov and Galakhov). From G. V. Kukolev (1966). 'Chemistry of Silicon and Physical Chemistry of Silicates'. Vysshaya Shkola, Moscow.

from which we note that the only solid phase that remains stable in these refractories at a temperature above 1 600°C is mullite, with the formula $3Al_2O_3.2SiO_2$. Mullite contains 72% alumina and 28% silica. It crystallises in the rhombic form as needles, prisms and fibres. It is the most important single compound in the production and use of fireclay refractories and also certain other types of refractory material.

Fireclays contain many impurities as well as silica and alumina and these may reduce the temperature at which melt appears to 1 345°C.

The alumina content roughly determines the 'quality' of a firebrick because the amount of solid phase in aluminosilicate refractories in service at high temperatures depends on the alumina content. After the mullite has crystallised (the precise temperature can be read off the phase diagram) the remaining silica is changed into a liquid phase, as are also the remaining fluxes.

Referring to the phase diagram for alumina and silica it is seen that in the case of kaolin and fireclay refractories, as the alumina content increases the concentration of liquid phase at a given temperature gradually falls, and one may be tempted to assume that (as mentioned above) the quality of the firebrick will increase with the alumina concentration. However, this is not borne out in practice because of the presence of impurities such as potassium, sodium, calcium and magnesium which make the system other than binary.

It is necessary in drawing conclusions from phase diagrams for fireclays and firebricks to consider other systems such as the $Al_2O_3$–$SiO_2$–$CaO$ system, the $Al_2O_3$–$SiO_2$–$Fe_2O_3$ system, etc. From these, fuller information about the real behaviour of fireclays and firebricks in service and during firing will be possible.


## MAIN PROPERTIES OF FIREBRICK

The chemical composition and other properties of fireclays used to make firebrick vary very widely. The refractoriness (PCE) of firebrick, kaolin and siliceous firebrick comes within the range 1 570–1 770°C but it is largely determined by the chemical composition and the phase composition of the product after firing. Firebricks based on china clay or kaolin raw materials generally have a higher refractoriness than fireclay brick (up to 1 750°C). The alumina concentration in firebrick ranges from 28–46%. Aluminous firebrick is normally made from a refractory material containing 38–45% $Al_2O_3$. Above that concentration the material may be known as a high-alumina product.

The impurities of firebrick are also extremely variable and may include quite large amounts of iron oxides, alkalis, alkaline earths, and occasionally transition metals such as manganese. The titanium dioxide is frequently classified with the $Al_2O_3$ in reporting chemical compositions.

*Refractoriness under load* depends on the chemical composition and the physical structure of the material. Fluxing impurities have a critical bearing on the temperature and manner in which the refractory fails under load. The most critical impurities for a brick containing about 40% $Al_2O_3$, for instance, are the alkalis, the alkaline earths and the iron and manganese oxides.

The volume stability of firebrick depends on the temperature and time for which it remains in the hot zones of furnaces. When soaked for long periods at temperatures substantially below their maximum service point, firebricks may exhibit considerable shrinkage and even drop out of the furnace lining; when severely overheated firebricks will lose their structural strength, deform, twist and fuse. Firebricks containing chamotte that has been calcined at very high temperatures are the most stable in service. As mentioned above, siliceous firebrick during service may have a much lower shrinkage than ordinary firebrick and may even show an expansion owing to the crystalline inversion of the silica which is present in large proportions. Such siliceous materials are suitable for furnace linings provided the furnace charge is not highly basic, a condition which would make it capable of reacting very vigorously with the acid lining.

*Spalling resistance* of firebrick is fairly high with high-grade products but much depends on the composition, the fabrication method and the final structure of the brick. The relationship between spalling resistance and structure is discussed in Chapter 2.

One method of improving the spalling resistance of firebrick is to increase the concentration of chamotte. This reduces the amount of glassy phase, which is the main source of spalling cracks. It is generally considered that the more porous a firebrick the greater its spalling resistance, although no direct relationship has been detected between the spalling resistance and the porosity of firebrick.

Firing firebrick to a very dense structure such as that required in the production of ladle bricks which have to withstand the penetrative action of molten metal and slags, tends to impair the spalling resistance, since hard firing may increase the amount of liquid formed in the brick, impairing the strong, granular structure and leading to the formation of a structure similar to that of porcelain. This dense, impermeable structure is very susceptible to spalling attack.

*The slag resistance* of firebrick depends mainly on the density and porosity. The factors involved are discussed in Chapter 2. The aim in making slag-resistant firebrick is to close all open pores, eliminate fine and large cracks and cavities formed during fabrication and to present a monolithic, impermeable face to the slag attack. Here again there is no proven relationship between the slag resistance and the total porosity and several factors are relevant; for instance, whether or not the pores are communicating (*see* discussion under 'permeability' in Chapter 2).

Although the chemical composition of firebrick is not closely related to slag resistance, it is generally accepted that any product containing more than $40\%$ $Al_2O_3$ has a greater slag resistance to the same type of slag than materials with lower $Al_2O_3$ concentrations. As might be expected, siliceous products resist the action of acid slags better than basic products, and the resistance of these refractories to basic slags is poor.

Table 6.1 shows the properties of various fireclay refractories made in various countries of the world.

## USES OF FIRECLAY REFRACTORIES

Owing to its relative cheapness and the widespread location of the raw materials used to manufacture firebrick, this material finds uses in most furnaces, kilns, stoves, and steam raising equipment. Firebrick is

TABLE 6.1

*Properties of Some Common Firebricks Made in Various Countries*

| Property / Country | $Al_2O_3$ (%) | PCE (°C) | RUL[a] 2 kg/cm² (°C) | Porosity (%) | After-contraction (−) or expansion (+)[b] (%) | Compressive strength (kg/cm²) |
|---|---|---|---|---|---|---|
| UK | 29–43 | 1 600–1 720 | 1 200–1 400 | 10–23 | −0·2–0·9 | 150–480 |
| USA | 28–44 | 1 530–1 750 | 1 150–1 410 | 9–21 | −0·1–+0·4 | — |
| USSR | 28–41 | 1 690–1 730 | 1 305–1 390 | 14–22 | −0·2–1·7 | 140–540 |
| W. Germany | 22–35 | 1 610–1 710 | 1 230–1 430 | 11–26 | — | 160–388 |
| France | 26–41·8 | 1 680–1 770 | 1 140–1 300 | 21–30 | — | — |

[a] RUL—refractoriness under load.
[b] Tested for 2 h at 1 410°C in oxidising conditions.

the commonest form of refractory material. It is used extensively in the iron and steel industry; non-ferrous metallurgy; the glass industry; for building pottery kilns; for the cement industry and in all other heat-processing industries. In the manufacture of steel, firebrick is employed for blast furnaces; regenerator stoves and iron ladles; for building certain open-hearth furnace parts and in steel casting (ladles). In short, to the average person the mention of the words refractory or heat-resistant material immediately brings to mind the ubiquitous firebrick. Although it is fast losing ground to more sophisticated and expensive products, there is little doubt that the firebrick will remain popular for many demanding and undemanding uses for many years to come.

*Chapter 7*

# High-alumina Refractories

Aluminosilicate refractories containing more than 45% alumina are generally termed high-alumina materials. The alumina concentration in high-alumina refractories ranges from 45 to 100% (pure alumina).

These refractories are made, by a variety of processes, from bauxite, sillimanite (kyanite and andalusite are similar to sillimanite), diaspore, mullite, fused and calcined alumina, and from aluminous fireclays and kaolins (Fig. 7.1). High-alumina insulating products are made from kaolin mixed with calcined alumina, and also from bubble alumina—a product obtained by fusing alumina and casting it to yield hollow spheres of various diameters.

Perhaps the simplest method of making high-alumina refractories is to grind raw bauxite, blend it with a small amount of plastic clay and press this into the desired shapes. These are then fired in a ceramic kiln. Sillimanite minerals can also be treated in the same way.

Bauxite found in nature may contain several forms of hydrated alumina, and iron and silica impurities. The hydrated aluminas are diaspore (monohydrate) $Al_2O_3.H_2O$; boehmite (monohydrate) $Al_2O_3.H_2O$; bayerite (trihydrate) $Al_2O_3.3H_2O$; and gibbsite (trihydrate) $Al_2O_3.3H_2O$.

When heated to 900–1 100°C these hydrated forms lose their water of crystallisation to yield $\gamma$-alumina (spinel form). With further heating this form changes to the more stable $\alpha$-form (corundum). There is also another form known as $\beta$-alumina (*see* below).

Corundum refractories and ceramics constitute an important class of materials because of their high refractoriness, chemical resistance, and electrical properties.

## CORUNDUM (ALUMINA) REFRACTORIES

Refractories containing more than 95% $Al_2O_3$ are classified as corundum ceramics. Their properties and uses will depend on the additives making up the remaining 5%. The raw material is usually one of the anhydrous types of alumina used in the form of calcined or fused alumina.

89

FIG. 7.1.  Kaolins may be briquetted, fired, crushed and ground to produce aluminous chamottes. Here cars of china clay are awaiting calcination in a tunnel kiln. *Photo:* Courtesy English China Clays Group.

The main crystal forms and some properties of pure alumina are shown in Table 7.1.

TABLE 7.1

*Crystal Forms of Alumina*

| Alumina form | Crystal system | Optical properties $N_o$ | $N_e$ | Specific gravity |
|---|---|---|---|---|
| $\alpha$ | Trigonal | 1·765 | 1·757 | 3·99 |
| $\beta$ | Hexagonal | 1·66–68 | 1·63–65 | 3·31 |
| $\gamma$ | Cubic | 1·73 | — | 3·65 |

Some other properties of alumina are: melting point 2 050°C; Mohs hardness 9; Rockwell hardness 90; density 3·98–4·01 (the test methods affect the results of density determination).

The $\beta$-alumina is not a pure form of alumina but may contain a variety of alkalis. Incorporating alkali ions in the alumina crystal structure yields a range of compounds with the formulae: $MeO.6Al_2O_3$ and $MeO.12Al_2O_3$, where Me may be BaO, CaO, or SrO, and $Me_2O$ may be $Na_2O$, $K_2O$, or $Li_2O$. The concentration of alkali may be 8–10%. When these $\beta$-forms are heated above 1 600–1 700°C the alkali is removed by volatilisation as the crystal lattice breaks down.

**Calcined alumina**

Calcined alumina is the product obtained during the extraction of aluminium by the Bayer process. The technique is to decompose the starting aluminous raw materials with a solution of caustic soda to form sodium aluminate in solution. The impurities remain in the undissolved state and are removed as a residue. The solution of aluminate is purified, after which pure aluminium hydroxide can be precipitated and calcined, usually in rotary kilns at temperatures of 1 100–1 220°C, to yield a uniform white powder (calcined alumina).

In the production of alumina refractories, this material, after blending with other ingredients, is usually fired at much higher temperatures, during which it is converted into the $\alpha$-form. The bulk density of the calcined alumina alters with the firing temperature. For example, in the original form after treatment by the above mentioned production method, the bulk density is 0·86 g/cm$^3$, when fired again at 1 500°C it rises to 1·2 g/cm$^3$ and when fired at 1 760°C to about 1·6 g/cm$^3$.

**Electrofused corundum**

This is produced by fusing calcined alumina in electric arc furnaces or by using bauxite and the same process. The alumina concentration in white electrofused corundum is 98% or more, and in bauxitic types 91–95% (impurities are silica and ferric oxide).

**Fusion-cast alumina**
Originally electrofused cast refractories were produced by casting molten bauxite into graphite moulds to yield a material containing about 75% alumina, being made up of mullite and corundum in a vitreous matrix. Today mullite–corundum castings with higher alumina concentrations and zirconia–corundum castings are commonly made, especially for use in glass furnaces. The batch materials are normally melted above 2 000°C and therefore electric furnaces must be used. The process of fusion casting is described in greater detail in Chapter 14.

**Sintered alumina**
The sintering of alumina is of more interest to producers of alumina ceramics, particularly in the development of electrical properties and abrasion resistance. When alumina is sintered under certain conditions it becomes gas impermeable due to densification which normally involves the formation of a vitreous phase and grain growth. High-density products can be obtained by sintering alumina with small quantities of magnesium oxide.

## MULLITE REFRACTORIES

A distinction should be made between the mineral mullite and the materials produced commercially and marketed as mullite refractories. The mineral mullite is a compound of alumina and silica, with the formula $3Al_2O_3.2SiO_2$. Because of its high-temperature stability and other valuable ceramic properties, mullite is encouraged to develop in many types of ceramics and refractories. The ways in which mullite formation is fostered are: by selecting the appropriate chemical compositions, by heat treatment, and by the use of mineralising additives in the batches (*see* below). The mineral mullite may appear in varying amounts in firebricks, in high-temperature porcelains, in fusion-cast refractories and even in some types of earthenware, stoneware and common bricks. The chemical composition of the batch and the firing temperatures determine the concentrations of mullite in these products.

*Synthetic mullite* is made by firing bauxite (or some other form of alumina) with quartz sand or kaolin. The main crystal phases of mullite and mullite–alumina refractories are mullite ($3Al_2O_3.SiO_2$) and $\alpha$-$Al_2O_3$. By varying the ratios of synthetic mullite and alumina in the batch it is possible to produce refractories with alumina concentrations from 56 to 100%.

The manner in which the phases in the alumina–silica system react when heated is, of course, critical for the service of high-alumina refractories. The system has been thoroughly researched and some difference of opinion has existed about the quantities and characteristics of the various phases and their behaviour during heat treatment. It now seems clear that when alumina concentrations rise above 72%, corundum also exists in the system as well as mullite. At 1 850°C these two crystal phases react with each other to yield a eutectic, causing the system to melt congruently. The composition of this eutectic is 77·2% $Al_2O_3$ and 22·8% $SiO_2$. In the compositional region between $3Al_2O_3.2SiO_2$ and this eutectic, mullite forms solid solutions with the alumina. As we increase the alumina content of the system above 77·2% we find mullite coexisting with alumina right up to 100% $Al_2O_3$ which melts at 2 050°C.

This information has largely been read off the phase diagram for the alumina–silica system, but it might be added that practical experiments and industrial experience support the predictions about the behaviour of high-alumina (mullite) refractories. This is in spite of the fact that commercial mullite refractories contain 4–7% fluxes, calculated as alkali oxides and alkaline–earth oxides (for example, soda, calcium oxide, etc.) which may reduce the sintering temperature of the materials to 1 400–1 450°C.

### Classification of mullite refractories

From the above it is apparent that chemical composition is the decisive factor in the proportions of mullite and other phases in high-alumina refractories and ceramics. It is possible to classify these refractories into the following three groups:

(a) mullite–siliceous containing 45–70% alumina;
(b) mullite–alumina (70–95% alumina);
(c) alumina refractories containing 95–100% $Al_2O_3$.

In general refractories containing less than 70% $Al_2O_3$ contain mullite and various quantities of glassy phase which may amount to 40–50%. As the alumina content rises to about 78% the quantity of glassy phase (providing the batch contains no fluxes) diminishes to a minimum, while that of the mullite increases. The varying proportions of the mineral phases obtainable in mullite–alumina refractories explains the great range of refractory products that can be produced from this system.

### Methods of producing mullite

The following methods of synthesising mullite are used industrially or on a laboratory scale.

(1) *Crystallisation from fusions*, in which suitable batches are melted at about 1 700°C, using chromium oxide and sodium tungstate as mineralising agents. It is found also that adding small quantities of beryllia, lithium fluoride, or zinc oxide reduces the temperature at which the mullite develops, to about 1 250–1 400°C. The technique of adding other mineralisers such as boron, manganese or titanium is said to have no

great effect on mullite formation, while adding trivalent iron and nickel tends to retard mullite formation. It is possible to induce mullite formation at 900°C in batches containing lithium fluoride, and at temperatures as low as 610°C when aluminium fluoride is added. Additions of beryllium, magnesium and calcium oxides in quantities of about 5% yield mullite with 90% of the theoretical density, and the mineraliser affects the shape of the mullite crystals.

TABLE 7.2

*Effect of Some Mineralising Additives on the Sintering Temperature of Synthetic Mullite*

| Additive | % by wt. | Sinter- ing temp. °C | Sintering range | Reduc- tion sintering temp. °C |
|---|---|---|---|---|
| None | — | 1 590 | 100–150 | — |
| MgO | 1 | 1 510 | 110 | 80 |
|  | 2 | 1 450 | 70 | 140 |
| CaO | 1 | 1 520 | 110 | 70 |
|  | 2 | 1 470 | 70 | 120 |
| MgO + CaO | 2 | 1 410 | 70 | 180 |
| MnO | 1 | 1 520 | 110 | 70 |
|  | 2 | 1 470 | 100 | 120 |

(2) *Crystallisation from the gas phase.* American patents cover processes in which mixtures of silicon, sulphur and aluminium heated in an atmosphere containing 1% hydrogen at 800–1 200°C yield mullite fibres by crystallisation from the vapour phase. Adding small quantities of aluminium fluoride to the vapour state accelerates the reaction.

(3) *Flame spraying.* Mullite possessing needle-like structures can be obtained by spraying onto substrata compositions of Si–Al fusions in an atmosphere of oxygen at a pressure of 10–15 mm Hg with a direct current between the cathode of Si–Al alloy and the substratum.

(4) *Solid phase reactions,* in which mullite is developed by sintering alumina–silica mixtures below the melting temperature of the mixture, have been used. The process involves the formation of an intermediate crystal phase. An example of solid phase mullite formation is that taking place in porcelain. In porcelain the mullite exists as flakes formed from the kaolin by solid-phase reactions, and needle-like crystals develop as a result of the solution of aluminosilicates in the melt, followed by recrystallisation.

(5) *Coprecipitation of alumina and silica gels* is another technique for

producing mullite at about 1 140°C, with the mullite being detectable by X-ray diffraction. (The preceding is reported as a summary of the relevant research in 'Principles of Solid State Chemistry', P. P. Budnikov and A. M. Ginstling, Applied Science, London, 1968.)

Table 7.2 indicates the effect of some mineralising additives on the sintering temperature of synthetic mullite.


## MANUFACTURING METHODS OF MULLITE AND HIGH-ALUMINA REFRACTORIES

As mentioned above the raw materials for making high-alumina refractories consist of a range of minerals mined naturally and synthesised by heat processing bauxite, clays and other compounds. They include andalusite, kyanite, kaolins, fireclays and processed alumina. The sillimanite group of minerals, which includes andalusite and kyanite, has the general formula $Al_2O_3.SiO_2$, and contains roughly $62.9\%$ $Al_2O_3$ and $37.1\%$ $SiO_2$. These can be used to obtain refractories with alumina concentrations of not more than $60\%$. By adding calcined alumina and other high-alumina minerals it is possible of course to increase this concentration.

Sillimanite minerals when heated are converted into mullite according to the following reaction:

$$3(Al_2O_3.SiO_2) \rightarrow 3Al_2O_3.2SiO_2 + SiO_2$$

The reaction occurs at temperatures between 1 300–1 500°C but as in the case of mullite formation in the alumina–silica system, in general the temperatures of the reaction depend very much on the presence of mineralisers (*see* above). In theory when sillimanite is heated the mullite yield should be $87.5\%$ and the quantity of free silica (cristobalite) $12.5\%$.

The mullite in commercial mullite refractories may be developed directly in the products by firing the raw materials, or it may be incorporated in the form of synthetic mullite. For high quality refractories the second method is preferred because the first technique is not subject to the close control of such properties as shrinkage and shape tolerance.

High-alumina products containing various concentrations of alumina can be made by the conventional ceramic process of blending the granular grog materials and clay, followed by pressing and firing. Other techniques involve fusing the raw materials and casting into graphite and other moulds to produce high density 'fusion-cast' products. The main feature of producing high-alumina refractories is the very high firing temperatures (1 500–1 680°C).

## PROPERTIES OF HIGH-ALUMINA CERAMICS

In addition to the high fusion point of mullite-type refractories, other important properties are their high resistance to thermal shock (spalling), due to the low concentrations of glass compared with fireclay types, and their high refractoriness under load, due to the higher degree of solid–solid bonding in the product.

The compressive strength of high-alumina refractories may be 1 200–30 000 kg/cm$^2$, the bending strength about 1 500–4 000 kg/cm$^2$, and the tensile strength around 2 200 kg/cm$^2$.

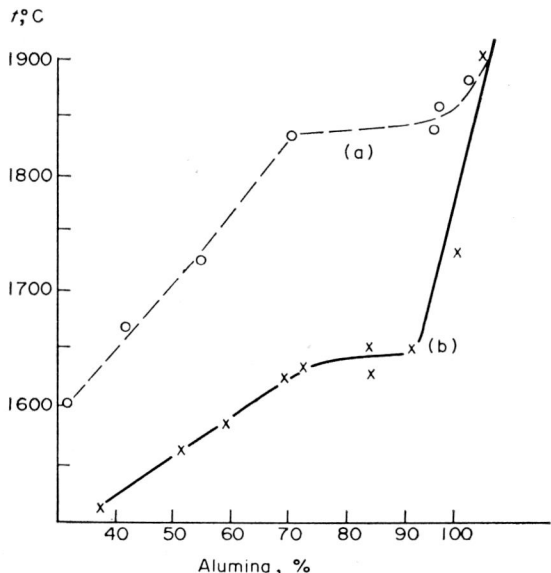

FIG. 7.2.    Effect of alumina content on the refractoriness under load (2 kg/cm²) of high-alumina brick. (a) 40% sag; (b) RUL.

In addition to the alumina concentrations the factors governing the mechanical and other strength properties of these refractories include the manufacturing technology such as grain size distribution, firing temperature, and the presence of mineralising additives as discussed above.

The refractoriness itself of course is dependent on the concentration of alumina, but is usually about 60–90° below the theoretical melting point (liquidus temperature) on the phase diagram for the alumina–silica system. Products containing 45–60% alumina have fusion points of around 1 750–1 820°C, those containing 70–95% alumina fusion points of 1 780–1 850°C, and the alumina products (above 95% Al$_2$O$_3$) fuse at 1 900–2 000°C. The relationship between alumina concentrations and the refractoriness under load of some mullite sintered refractories is shown in Fig. 7.2.

**Slag and molten metal resistance**
Raising the alumina concentration of aluminous refractories normally causes an increase in their chemical resistance in regard to slags and molten metals (but *see* Chapter 19, under *ladle brick*). The breakdown of mullite and high-alumina refractories by metallurgical slags and fluxes is due to alterations in the phase compositions in the silica–alumina system, resulting from the reactions between the slags and the refractories. It is worth noticing the difference in the behaviour of alumina (corundum) and mullite towards basic slags. Corundum is found to have a greater slag resistance than mullite. However, because of the 'acid' action of alumina in hot chemical reactions, even alumina refractories have a much lower resistance to basic slags than is shown by basic refractories such as magnesite, etc.

## USES OF ALUMINA REFRACTORIES

The number of applications of high-alumina refractories is very great. It includes the hearths and shafts of blast furnaces; stoves of blast furnaces; ceramic kilns; cement kilns; glass tanks, and furnaces and crucibles used for melting a wide range of metals including rare and noble metals. High-alumina porcelains are also used in the chemical industry, for making sparking plug insulators and for various special high temperature applications in aerospace research.

*Chapter 8*

# Silica Brick (Dinas)

The use of silica as a refractory has declined rapidly in recent years owing mainly to the application of oxygen in steel making, requiring refractories with higher fusing points, and also the development of new types of furnace and techniques, *e.g.* in gas making. However, no account of refractories would be complete without a description of silica.

Silica brick, or as it is known on the Continent of Europe, *dinas*, is a refractory material containing at least 93 % $SiO_2$. The raw materials used are quartz rocks, and lime or other bonding agent, blended and fired hard enough to make sure that the quartz is converted into cristobalite and tridymite to produce a satisfactory commercial refractory.

Judging from its composition, silica should be one of the simplest refractories to understand since it is made up mainly of $SiO_2$. However, the complexities of the changes that occur when it is heated either in the kiln used to fire it or in service, are such that even today arguments continue on the various crystal forms in which the mineral can exist: about two dozen forms of silica have been described that may either exist or are thought to exist in various states of stability.

The standard English-language book on silica and its crystalline forms is that by Sosman (1927). Because of recent research and the discovery of entirely new forms of silica such as silica W, compacted vitreous silica, piezo-vitreous silica, coesite, keatite and stishovite, much of Sosman's earlier account has been revised (by Sosman himself among others). A more recent, full-length treatment is contained in the book by Kukolev (1966). For the very latest information on quartz and silica reference should be made to the current ceramic and refractories literature.

The outstanding property of silica brick is that it does not begin to soften under high loads until its fusion point is approached. This behaviour contrasts with that of many other refractories; for example alumino-silicate materials, which begin to fuse and creep at temperatures considerably lower than their fusion points.

Other advantageous properties of silica refractories include flux and slag resistance, volume stability beyond certain critical temperatures, and high spalling resistance. Provided care is taken in warming up silica structures, for instance up to 600°C, or up to 300°C for hard fired materials, the resulting lining will give excellent service for many years.

## ALUMINA IN SILICA

*The alumina content of silica refractories is critical.* By beneficiating quartzites and other silica raw materials to reduce the $Al_2O_3$ content to a minimum, it is possible to improve the refractoriness and refractoriness-under-load of the resulting silica brick.

Some idea of the effect of alumina on the refractoriness of silica raw materials can be obtained from the following figures: top grade dinas raw material contains less than 1·3% alumina and less than 0·5% CaO and has a PCE of 1 770°C. Medium grade raw material contains about 1·5% alumina and about 1% CaO with a PCE of 1 750°C.

When the silica batch is blended various other substances are added, including calcium and iron mineralising agents, and so the total $SiO_2$ concentration becomes less than that in the starting raw materials. The composition and properties of raw quartzites are frequently improved by washing to remove impurities.

Quartzites, which make up the majority of silica raw materials, are very hard and dense materials with a Mohs hardness of 7. The porosity is about 0·2–3%. Other types of quartzites, such as those bonded with various calcarious impurities, have porosities up to 12%.

## SILICA CRYSTAL INVERSIONS

As mentioned above, the crystal changes taking place when quartzites are fired are critical in refractories technology. All quartzites expand in volume when heated owing to polymorphic transformations of the quartz lattice. The density may drop during firing from 2·65 to 2·4. The material, when fired, is embrittled and the porosity is also increased. The velocity of the transformation and the extent to which raw quartzites are broken up by heat is important. In the British pottery industry, for instance, quartz pebbles are calcined in shaft kilns in order to effect certain crystalline changes and also to render the material softer for subsequent ballmilling.

The main change taking place in quartz in the initial stages occurs at a temperature of 573°C and is known as the $\alpha$-$\beta$ quartz inversion. The $\alpha$-quartz (low-temperature form) which is stable at room temperature, is the form used in dinas manufacture. The $\beta$-quartz (high-temperature form) is not found naturally because it is stable only between temperatures of 573 and 870°C. At the latter temperature (870°C) the $\beta$-quartz changes into $\beta$-tridymite, but only when certain mineralisers are present in the system and provided the quartz is finely ground. Certain additives such as sodium tungstate are used to effect the conversion.

Again, $\beta$-tridymite is not found naturally since it is stable only between 870 and 1 470°C. If the temperature of tridymite rises above 1 470°C it is changed into $\beta$-cristobalite.

At the time of writing considerable controversy exists over tridymite as a valid silica phase. Information has appeared about new series of inversions in tridymites and no satisfactory clarification of the $\alpha$-$\beta$ cristobalite change has so far been published.

The $\beta$-cristobalite forms when $\beta$-tridymite is heated above 1 470°C. The change is very slow and the cristobalite remains stable to about 1 720°C. At this temperature the $\beta$-cristobalite may change into fused quartz without any volume change. Upon cooling $\beta$-cristobalite changes into $\alpha$-cristobalite. This form is metastable, existing under ordinary conditions in the non-equilibrium state.

Fused quartz is formed by subjecting the various forms of silica to very high temperatures. For example, $\beta$-cristobalite melts at 1 713°C. At temperatures below 1 470°C it becomes unstable down to 230°C when it changes to $\alpha$-cristobalite.


## FIRING SILICA BRICK

The changes that occur when a green silica brick is fired are of the utmost importance to an understanding of the nature of silica refractories. Both the manufacturer and the user need to understand these principles. The above information merely indicates the complexity of the changes taking place in dinas refractories. The main phase diagram of relevance to this topic is the $CaO$–$Al_2O_3$–$SiO_2$ diagram. Calcium and iron are added as the bonding materials, and alumina is usually already present in the quartz rock (see Fig. 8.1.) The other material added to the batch for making dinas refractories is some kind of temporary organic bond, such as sulphite lye or carboxymethyl cellulose, which burns out when the brick is fired.

The raw materials, such as precrushed and graded ganister or other quartzitic rock, the mineralisers and the temporary bond are moistened and blended and then pressed. After drying the products are fired. During the firing of dinas the following critical processes occur.

First the silica starts to react with the calcium oxide to produce calcium silicates which form solid solutions with the ferrous silicates. As the temperature rises the polymorphic inversions of the silica, discussed above, commence, and there is an expansion of the green brick. For example, when the $\alpha$-$\beta$ quartz change occurs at 573°C the change in the volume is 0·82%. Subsequently, when the $\beta$-quartz to $\beta$-tridymite change occurs at 870°C the change is 16·0% by volume, and when the $\beta$-quartz to $\beta$-cristobalite change occurs at 1 000°C the volume increase is 15·4%. The melt starts to form as the above reactions get under way. The quartz is dissolved in the melt, and under certain conditions it is precipitated from the saturated solution. This leads to recrystallisation of tridymite. Another important change undergone by the dinas brick in firing is in

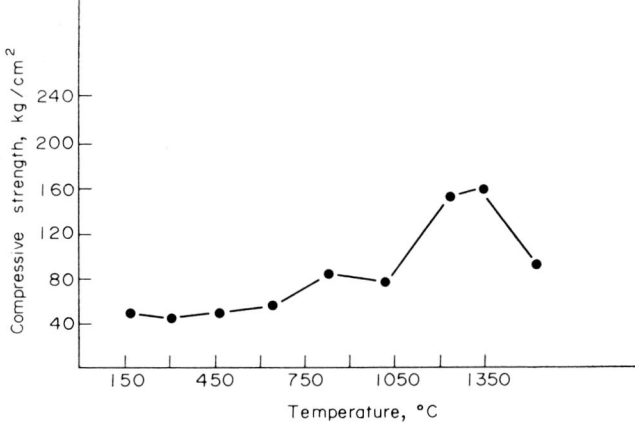

FIG. 8.1. Effect of temperature on the hot strength of silica brick, containing 1% FeO + 2% CaO as bond, during firing.

its mechanical strength. At certain critical temperatures it is very weak. These changes make it necessary to fire silica brick very carefully, otherwise the products either crack, split or crumble, and are then unsuitable for use.

Dinas brick sinters without any increase in bulk density, frequently with some embrittlement and even an increase in porosity. The product is strengthened during firing owing to the phase transformations (recrystallisation) and the reaction between the silica and the mineralisers such as calcium and iron compounds.

## DALE'S THEORY

The reactions taking place when silica brick is fired can be explained in accordance with Dale's theory (1927). This suggests that the lime and free quartz react at about 800°C and as the temperature rises the resulting lime–silica glass recrystallises and precipitates cristobalite between 1 250 and 1 300°C. Between 1 200 and 1 300°C the surfaces of the quartz grains are directly transformed into cristobalite and the rate at which this process takes place inwards rises with temperature increase. Cristobalite then dissolves in the lime–silica glass of the matrix. When the final firing temperature is approached the coarse grains are almost completely converted. The very fine quartz grains are completely converted and the silica concentration of the glass is rising rapidly owing to the solution of cristobalite. The final stage of the firing process consists in the precipitation of tridymite from the saturated glass.

The earlier processes are also important. At about 100–200°C the moisture is removed, and at 450°C the calcium hydroxide (slaked lime) begins to decompose rapidly, the decomposition terminating at 560°C. This reduces the mechanical strength of the green dinas. At 573°C the $\alpha$-$\beta$ quartz inversion occurs and is accompanied by critical internal stresses which may disrupt the brick if care is not taken with the firing process.

There is evidence suggesting that in the range 600–700°C the green brick shows some rise in strength due to reaction between calcium oxide and silica, resulting in the formation of $2CaO.SiO_2$ and a certain amount of free $CaO.SiO_2$.

Between 800 and 1 100°C pseudo-wollastonite is formed from the reaction between calcium silicates and the fine silica. At about 1 000°C the $\beta$-quartz changes into $\beta$-cristobalite and once again damaging internal stresses may develop and crack the products. At 1 100–1 350°C irreversible quartz transformations occur, producing silica forms with a low density. A liquid phase then appears and conditions develop for the crystallisation of tridymite which eventually forms the intergrowth framework of the silica brick. The products exhibit an increase in volume, the strength drops, and the internal stresses increase in magnitude.

Normal firing cycles for silica specify that the heating rate between 1 350 and 1 430°C must be used very carefully, since the crystal changes are very rapid in this range. The products must then be cooled and down to 400°C the cooling rate may be fairly quick. At this temperature and especially between 300 and 100°C special care must be taken since the $\beta$-$\alpha$ cristobalite change occurs with a reduction in volume of 2.8% and the tridymite conversions take place with a slight fall in volume (1.2%). Appreciable stresses may therefore develop, causing cracking if the cooling is done too fast.

The various forms of quartzitic raw materials used by different dinas brick producers usually mean that in practice a firing and cooling cycle is developed for the particular local raw material, and this cycle must be adhered to for consistent results.

Firing cycles normally involve the use of a slightly reducing atmosphere when the temperature is rising, particularly during soaking. There must be no sharp variations in temperature nor contact between flame and product. A thorough soaking at the final firing temperature is necessary to ensure satisfactory conversion. The total firing cycle for common silica brick lasts about 130 hours (periodic kilns) and about 200 hours for tunnel kilns.

In addition to the outstanding underload features of silica brick mentioned above, silica also maintains a constant volume with temperature change, above certain critical values. It has a high spalling resistance providing that certain conditions are met.

Silica is an acid refractory and so oxides of iron, calcium and other metals react with it and form complicated fusible silicates which may cause slagging. The hot surfaces of silica linings are found to form *icicles* which gradually flow down the walls of the furnace when fluxing materials are present. The slag resistance of silica can be increased by increasing its density but this may reduce the spalling resistance. A feature of silica brick is its tendency to become embrittled above 1 600°C due to further quartz transformations taking place. The density and strength are diminished and there is an increase in the porosity. The fault depends on the properties and the grain size of the original quartzite, especially the fractions less than 0·08 mm, the chemical composition of the mineralisers and the firing temperature used for the manufacturing process. The cure may be to add ferruginous mineralisers to the batch, use finer batch materials and burn the products to a lower density.

**Superduty silica brick**
This material (patented) contains less than 0·5 % $Al_2O_3$ + $TiO_2$ + alkalis. In Britain the term 'superduty silica' is reserved for a silica refractory containing not more than 0·5 % alumina and containing a total of 0·7 % $Al_2O_3$ + alkalis. The use of specially processed low-alumina raw materials to produce a very low-alumina silica brick enhances the working properties considerably and great care is taken in the industry to reduce the alumina content by as little as 0·25 % in order to obtain worthwhile improvements in performance. It is possible that the titanium oxide concentration of the brick has little critical affect on the refractory performance.

**Uses of dinas (silica brick)**
Various grades of silica brick have found extensive use in the iron and steel industry, glass melting, and in tunnel kilns for firing refractories and ceramics. The steel industry used to employ large quantities of dinas for the roofs of basic open-hearth furnaces, regenerator checkers and slag pockets. In the glass industry furnace parts made out of silica include the roofs, the doghouses, the burner ports and in some furnaces the regenerators. Kilns used for firing dinas have their roof walls and bases and other

parts built of silica brick. The firing zones, roofs and walls of kilns used for firing ceramics and fireclay refractories may also be made from silica brick.

## REFERENCES AND BIBLIOGRAPHY

Chesters, J. H. (1963). 'Steelplant Refractories.' United Steel Company, Sheffield, pp. 67–117.
Dale, A. J. (1927). *Trans. Brit. Ceram. Soc.* **203,** 217.
Ford, W. F. (1967). 'Effects of Heat on Ceramics'. Applied Science, London, pp. 74–102.
Kukolev, G. V. (1966). 'Chemistry of Silicon and Physical Chemistry of Silicates' (in Russian). *Vysshava Shkola*, Moscow (English translation in preparation).
Sosman, R. B. (1927). 'The Properties of Silica'. Chemical Catalogue Co., New York.
Worrall, W. E. (1967). 'Raw Materials'. Institute of Ceramics Textbooks, London.

*Chapter 9*

# Magnesite

Magnesite refractories are chemically basic materials containing at least 85% magnesium oxide and consisting chiefly of the mineral periclase, MgO. They are made from naturally occurring magnesite ($MgCO_3$) and from chemically processed magnesia produced from seawater. The principal *impurities* of these refractories are iron, calcium and silica, present in the starting materials as forsterite, magnesioferrite and monticellite.

Refractories based on magnesite take the form of fired and unfired bricks and special shapes and also powder used for ramming the linings of and for fettling steel furnaces. Magnesite powders are also used in refractory concretes.

Magnesium oxide melts at 2 800°C but impurities in commercial refractories impair this very high refractoriness. Nevertheless, magnesite arches have been used in steel melting at 2 000°C: a very high temperature for any refractory.

## PRODUCTION

The commercial production of magnesite refractories follows the pattern of most other refractories, that is, raw material preparation and particle size grading, fabrication by pressing, drying and sometimes firing. Unfired brick is used with a steel cladding or case.

The magnesite rock, $MgCO_3$, is calcined to remove the carbon dioxide, leaving magnesia. If the process is done at temperatures high enough to eliminate most of the shrinkage and to yield low-porosity grains, the product is known as 'dead burnt' magnesite. Temperature is not the only factor, however, and the presence of fluxes will yield a liquid phase in the calcined material, thus densifying the mass by vitrification.

Calcined magnesite (N.B: the term 'magnesite' is retained for the product of calcination even though strictly speaking it refers to magnesium carbonate) is prone to hydration. When it reacts with water magnesium hydroxide is produced. In manufacture and use hydration needs to be restricted and various techniques are used to achieve this. Dead burning reduces hydration very significantly. Unfortunately, the purer the raw magnesite the higher the dead-burning temperatures required to render the product immune to hydration.

In producing stable magnesia for refractories the aim is to make a dense grain, free of micropores. Seawater magnesia satisfies these requirements and refractories made from it are relatively free from hydration troubles. Hydration would cause the bricks to crack and soften in storage and use.

Various additives may be incorporated with the magnesite to accelerate dead burning, *e.g.* calcium ferrite. Alumina is also sometimes added to make a better bond.

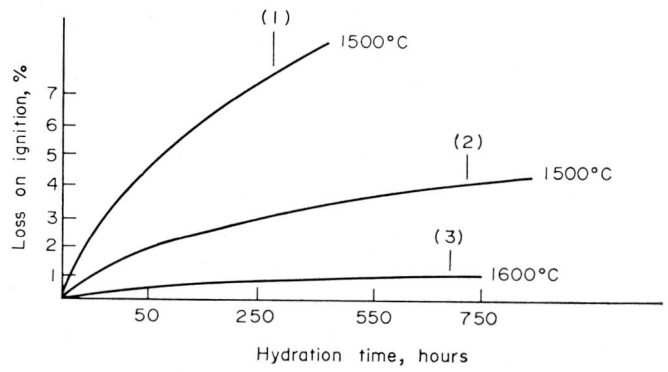

FIG. 9.1. How magnesite burnt at different temperatures hydrates. (1) Stored at 60°C; (2) at 40°C; (3) at 20°C.

### Sintered magnesite

This is the common raw material of the refractories producer. It is calcined in periodic or rotary kilns at 1 550–1 650°C, undergoing a shrinkage of about 25%. Dry or wet calcining methods are used, with oil, gas or solid fuel. Chemical and grain-size compositions, furnace atmosphere, additives, and of course the firing temperature and time all affect the process and the quality of the magnesite.

Figure 9.1 shows the way in which magnesite hydrates as a function of time and temperature of calcination. Figure 9.2 illustrates apparatus used to study the action of various gases on materials at high temperatures.

### Phase systems and solid solutions

Magnesia forms a continuous series of solid solutions (magnesiowustites) with ferrous oxide, FeO. A knowledge of the relevant phase diagrams will enable us to predict roughly the behaviour of magnesite in service. Even with 50% FeO in the magnesite, the fusion point remains high.

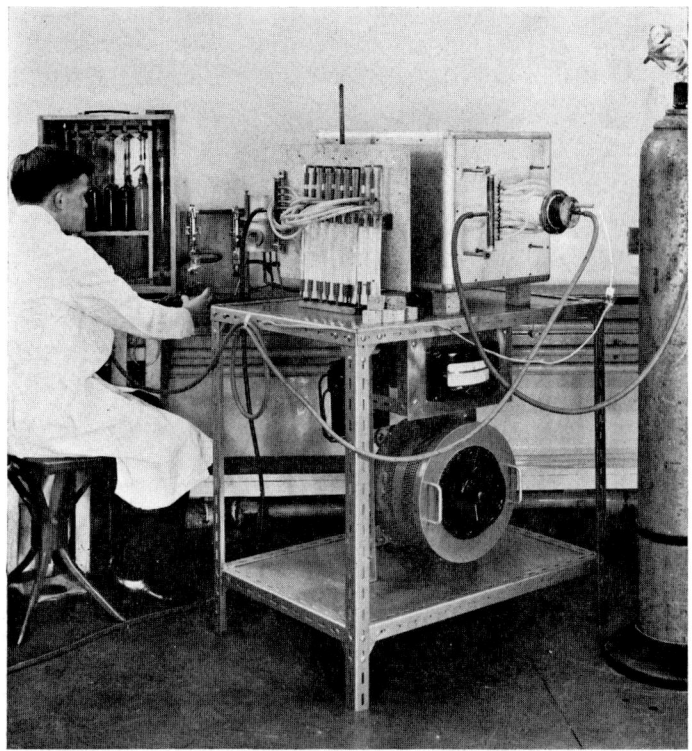

FIG. 9.2.   Researcher at the British Ceramic Research Association studying the action of various gases at high temperatures on refractory materials. *Photo:* Courtesy British Ceramic Research Association.

Under oxidising conditions $Fe_2O_3$ and magnesite produce magnesio-ferrite which is mutually soluble with periclase (MgO) at high temperatures. Thus, adding FeO to magnesite accelerates sintering and helps the highly refractory periclase mineral to crystallise. This is partly the reason why magnesite refractories are highly resistant to iron oxides and ferruginous slags and why these refractories are so important in steel making.

On the other hand, magnesite refractories are unsuitable for high-temperature service in contact with aluminosilicates and highly aluminous slags, since a eutectic forms in the $MgO–Al_2O_3–SiO_2$ system, having a low melting point (1 345°C).

In the $MgO–CaO–SiO_2$ system a eutectic forms (30·6% CaO, 61·4% $SiO_2$, 8·0% MgO) which melts at 1 320°C. There are several other non-refractory compounds due to the simultaneous presence of calcium and silica.

### Fired and unfired magnesite bricks
The powdered magnesite is pressed to confer the necessary shape, but instead of firing in a ceramic kiln, as is usual to make the material hard and durable and to effect ceramic processes, steel cladding is often put round the pressings during fabrication. Alternatively the pressed brick can be laid into the furnace linings with loose steel plates and sheets.

Steel-clad magnesite shapes are very strong and can be clipped into the linings or roofs or welded into place.

Today most fired magnesite refractories are produced in tunnel kilns. Since the firing temperature is high and coincides with the point at which the material begins to squat under a high load the bricks are set low on the kiln car (1·5 m high). The kiln atmosphere is slightly oxidising to keep the iron in the ferric state and help sintering of the magnesite grains.

During firing the phase composition of magnesite brick does not radically alter. This is convenient in the use of unfired bricks because any serious changes taking place in firing would obviously occur (in the unfired brick) during the initial service in the furnace lining. Large volume changes, for example, could cause the lining to crack. Fortunately, magnesite undergoes no phase inversions such as occur in silica.

However, between 420 and 1 250°C the green strength of the magnesite brick, which is due to the hydrated CaO and MgO 'cements' (hydraulic bonds) is lost when these compounds are dehydrated by the kiln's heat. No liquid has yet formed at these temperatures and so the bricks are mechanically very weak. Cracking may occur if the firing cycle is incorrect. Once 1 450–1 500°C is reached the resulting liquid starts to bond the grains of magnesite and the firing can be speeded up. However, above 1 500°C there is a danger of squatting and deformation due to excessive liquid formation.

Magnesite batches are usually soured (aged) in the wet state or otherwise subjected to a degree of controlled hydration before pressing. This is particularly important for magnesite high in free lime or soft-fired batches of grain. Modern practice involves autoclaving with steam. This prevents cracking during firing.

### Fettling magnesite
Sintered magnesite powders are used extensively for repairing (fettling) the hearths of furnaces while they are still hot. The technique saves the time and the considerable cost of cooling and reheating the furnaces. The fettling of open hearth and electric furnaces is an important constituent of steel plant economics.

Magnesia is an excellent fettling material because it is resistant to ferruginous slags and particularly to ferrous oxide, as discussed above. It is prepared in various grades in terms of the grain size and chemical composition. Since the hearth is restored by throwing small quantities of fettling powder into the hot furnace, the grain-size distribution and the 'fluid' properties of the powder are critical in building up a strong hearth. Experienced operators suggest that the MgO content should not exceed 85% and the $SiO_2$ should be less than 4–6%. The $Fe_2O_3$ and CaO contents appear not to be critical.

It is common in open-hearth practice to mix 10–15% open hearth slag with the fettling magnesite. The final (surface) layer is usually different from the underlayers and consists of magnesite, dolomite, and iron ore. This gives a final fired phase composition of 67–78% periclase and 10–15% dicalcium ferrite.

When a furnace hearth is being fettled liquid and solid-phase sintering occur in the periclase crystals. The fettling grains, rapidly heated to 1 550–1 650°C, crack explosively. The outer zones are impregnated by iron, calcium and silicon oxides, and the periclase crystals expand causing the fragments to pack tightly together. Furthermore, a ferrosilicate liquid is formed aiding the hearth building process. The layer thus formed is made up of magnesiowustite crystals 0·025–0·3 mm in size, cemented by calcium ferrites and silicates. A newly fettled hearth will contain 45–49% MgO and 35–40% iron oxides. As soon as melting recommences the iron oxide content falls, and the $SiO_2$, MnO and CaO contents rise. The MgO content remains fairly steady.

## PROPERTIES

The properties of magnesite refractories depend on the concentration of silicate bond at the operating temperatures. In magnesite this is determined by the silica content of the brick. Monticellite bond compositions adversely affect the properties (*see* below), whereas forsterite bonds improve them.

Good quality magnesite usually results from a $CaO–SiO_2$ ratio of less than 2 with a minimum ferrite concentration, particularly if the furnaces lined with the refractory operate in oxidising and reducing conditions.

The refractoriness under load of ordinary magnesite is much lower than the PCE value (1 500–1 550°C compared with 2 000°C or above). The cold compressive strength is high. The volume changes during heating are negligible. Spalling resistance depends on the structure of the fired brick; it seems to be higher for seawater magnesite brick than for natural rock magnesite.

As already discussed, the *slag resistance* is very high particularly to lime and iron rich slags. Attack usually occurs through the bond, but the magnesite grain itself may be attacked and be removed by erosion.

When heated under loads of 2 kg/cm$_2$ (50 psi), magnesite behaves very much like silica brick. It shows no sign of failure until the softening point is reached; then its collapse is sudden owing to the rapid reduction in the viscosity of the bond of forsterite or monticellite when heated only slightly above their melting points. This behaviour contrasts strongly with that of firebrick (see Chapter 6).

The refractoriness under load depends on the fabrication pressure used in making the brick or shape, the fired composition and the bond concentration, that is, of aluminosilicates. Magnesite brick containing monticellite

TABLE 9.1

*Properties of Magnesite Refractories (Fired)*

| | |
|---|---|
| MgO content, % | 93–96% |
| CaO content, % | 1·3–2% |
| Refractoriness under load, 2 kg/cm$^2$, °C | 1 500–1 550$^a$ |
| Apparent porosity | 20–27% |
| Apparent density | 2·6–2·75 |
| Spalling resistance, cycles | 4–30+ |
| After-contraction (at 1 500–1 600°C) | 0·2–1·0% |
| Compressive strength, kg/cm$^2$ | 450–650 |

$^a$ With $Al_2O_3$ additions this may rise to 1 620–1 670°C.

as the bond, as it were, sticking together the highly refractory periclase grains, has a poor refractoriness under load (below 1 500°C). Brick with a forsterite bond is much better and has a refractoriness under load greater than 1 600°C, since forsterite, $2MgO.SiO_2$, has a fusion point of 1 900°C.

Sometimes alumina is added to the magnesite batch in order to foster the formation of spinel bond and by using very pure magnesite (less than 0·5% $SiO_2$) it is possible to make refractories with a refractoriness under load of greater than 1 700°C. Reducing the silicate concentration may also improve the spalling resistance.

Table 9.1 shows the general properties of magnesite refractories.

**Unfired magnesite**
The properties of unfired (steel cased) magnesite are not easily summarised. The useful properties, that is, the properties prevailing in service, may alter radically from those determined before use by laboratory test.

## BIBLIOGRAPHY

Richardson, H. M., Lester, M. *et al.* (1969). Effect of boric oxide on some properties of magnesia, *Trans. Brit. Ceram. Soc.* **68**, No. 1, 29–31.
Staron, J. (1969). Pores in magnesite refractories, *Tonindustrie-Ztg. und Keram. Runds.* **95**, No. 5, 165–9.

## Chapter 10

# Magnesite–Chromite Refractories

A distinction must be made between chrome–magnesite refractories and magnesite–chromite refractories, although it may be rather tentative (*see* Chapter 3). Chrome–magnesite materials usually contain 15–35% $Cr_2O_3$ and 42–50% MgO whereas magnesite–chromite refractories contain at least 60% MgO and 8–18% $Cr_2O_3$.

Both classes are widely used in the steel industry, for which producers make highly spalling-resistant and slag-resistant bricks. They have replaced silica brick in most steel furnace roofs. For example, by 1960 more than

TABLE 10.1

*Compositions*

| | |
|---|---|
| Chrome brick | $Cr_2O_3$ + at least 10% magnesia |
| Chrome–magnesite | 70% chrome, 30% magnesite, or |
| | 60% chrome, 40% magnesite |
| Magnesite–chromite | Less than 50% chrome ore and more than 50% |
| | magnesite fusion cast (main constituents) |
| Fusion cast (main constituents) | 60% MgO, 20% $Cr_2O_3$, 8% $Al_2O_3$ |

95% of all Russian melted steel was being made in steel furnaces with chrome–magnesite roofs. In Russia the two grades are differentiated by calling the chrome–magnesite 'ordinary', and the magnesite–chromite 'spalling-resistant'. Chromium–magnesium refractories are high-duty materials and will withstand very high temperatures and corrosive conditions.

Specialised materials based on these compositions are being developed with even better properties (Table 10.1).

## PHYSICOCHEMICAL REACTIONS

When chrome ore (which is a mixture of chrome spinels) is blended with magnesite and fired by the refractories manufacturer in order to make

111

bricks, the magnesite reacts with the chromite, converting the fusible impurities present in the chromite into refractory compounds. The reactions are complicated and the full picture not clear but what takes place is roughly as follows.

The serpentine $(3MgO.2SiO_2.2H_2O)$ in the chrome ore decomposes at about 1 100°C to form $MgO.SiO_2$ at 1 350–1 400°C, and also forsterite $(2MgO.SiO_2)$. The fusion points of these compounds are 1 557 and 1 890°C respectively. The reaction is:

$$3MgO.2SiO_2.2H_2O \rightarrow 2MgO.SiO_2 + MgO.SiO_2 + 2H_2O$$

If an excess of magnesite is present, as is the case in commercial production, the following reaction occurs:

$$MgO.SiO_2 + MgO \rightarrow 2MgO.SiO_2$$

thus forming more forsterite and enhancing the refractoriness of the materials obtained after firing.

The net effect is that much of the impurity content of the chrome ore in the refractory batch is changed to the desirable forsterite.

Another essential reaction that occurs when magnesite–chrome mixtures are fired is that the ferrous oxide (FeO) of the chrome is replaced by the magnesium oxide (MgO) from the magnesite. There is a slight shrinkage at the reaction temperature of about 1 450°C, which depends on the grain sizes of the reactants.

The FeO replaced by the MgO separates on the surface of the chromite grains and is oxidised to ferric oxide $(Fe_2O_3)$ with a resulting increase in volume causing the mass to be embrittled. The process is accelerated with temperature rise (this is one of the reasons for the bursting of chrome–magnesite refractories).

## IRON-OXIDE BURSTING

This is a serious fault in chrome–magnesite refractories and manifests itself in the form of 'cauliflower' growths on the working surfaces of the linings coming in contact with iron slags. Various explanations have been offered, all of which take into account the marked increase in volume in the brick due to the effect of iron oxide. One school of thought postulates a volume increase due to the unequal diffusion rates of chromium oxide and iron oxides, causing the material to become porous. Another explanation is based on the solid solution of magnesite $(Fe_3O_4)$ in the spinels of the chrome ore. This hypothesis was subsequently contradicted because it did not account for the marked bursting expansion and was replaced by the view that the iron oxide entered the brick in the vapour form before reacting with the grains of chromite (spinel mixtures) to cause the sudden and disruptive expansion known as bursting. But even this explanation has since been questioned on the basis of experimental evidence.

The bursting of chrome ores when heated with magnesite tends to worsen with an increase in the $MgO:FeO$ and $Cr_2O_3:Al_2O_3$ ratios. The phenomenon is also more marked in fine-grained than in coarse-grained chrome-spinels probably because in the latter the presence of cracks and cavities compensate for the bursting expansion. Any measure that tends to reduce the diffusion processes taking place in the reaction zones on the periphery of the grains would help to eradicate iron-oxide bursting in chrome–magnesite refractories.

Research is continuing into this problem. Indeed the physicochemical principles of chrome–magnesite refractories technology are still being developed and the above notes merely indicate the complexity of the field.

It will be seen that when mixes of chrome ores and magnesia are fired the resulting reactions cause fundamental phase changes to occur, many of which are critical for the refractories producer and user. These changes depend on the chrome–magnesite ratios and their chemical composition, including the compositions of impurities; the particle sizes of the grade; the firing temperature and firing time and the cooling time. All this means that a wide variety of refractories is available, depending on the materials used and methods and the intended use.


## PRODUCTION

Chrome–magnesite refractories are made by the normal ceramic methods (blending graded grains, pressing and firing) and also by fusion casting. Unfired products are also produced and used in furnaces direct, a practice which is aided by the slight shrinkage of the green brick in firing. The compositions of unfired and fired brick are approximately the same and special care is taken with the content of fine fractions which increase shrinkage.

The batch is normally blended in runnermills, treated with a green (temporary) bond such as sulphite lye, pressed at 1 000–1 500 kg/cm² and then fired in tunnel or periodic kilns at 1 600–1 750°C, depending on the raw materials used.

The purpose of firing chrome–magnesite and magnesite–chromite refractories is to convert the batch components into stable, highly refractory spinels and to bond the silica into forsterite and tricalcium silicate. The physical–chemical changes occurring when batches are heated were mentioned above.

## SERVICE CONDITIONS

In service, chrome–magnesite develops a zoned structure. The hot zone is dense and if cooled and broken exhibits a metallic shine. It consists of complicated spinels and odd grains of periclase and chromite. The chemical composition of the working zone is found to have altered considerably from the original composition. The iron oxide content rises from above 5–10% to 30–45%. The chromium and magnesium oxide contents fall markedly, while the calcium and silica concentrations in the upper zone of the working face are increased.

As in all hot-face refractories, stresses develop in the lining, causing cracking and spalling. Failure results mainly from spalling and flaking and rarely because of fusion. The magnesite–chrome material may, however, show signs of fusion at very high temperatures. A serious fault is iron-oxide bursting, (*see* above).

## PROPERTIES AND USES

Chrome–magnesite refractories are made in a wide range of qualities, and are used for building the critical parts of high-temperature furnaces, (*see*

FIG. 10.1. Apparatus for determining the cold crushing strength of refractories such as fired magnesite–chromite refractories. *Photo:* Courtesy British Ceramic Research Association.

Fig. 10.1). These materials withstand corrosive slags and gases and have a high refractoriness under load when constituted for that purpose. The magnesite–chromite products (at least 60% MgO) are suitable for service at the highest temperatures and in contact with the most basic slags used in steel melting. They are used, for example, for making stopper tubes for ladles in which molten steel is vacuum treated: conditions which would severely affect the stability of any material. Magnesite–chromite usually has a better spalling resistance than chrome–magnesite.

Standard chrome-refractories do not react with silica brick up to 1 700°C and can therefore be used as the dividing layer for chromite–silica brick linings.

Magnesite–chromite has rapidly replaced silica brick (dinas) in the roofs of open hearths because of the following advantages:

1. Higher PCE; 2 300°C compared with 1 730°C.
2. Approximately the same initial softening under load (1 550–1 670°C).
3. The working temperature can be raised to 1 780°C compared with a maximum of 1 670°C for dinas.
4. Higher spalling resistance, leading to faster cycles for warming up furnaces.
5. Higher thermal conductivity, which means melting times can be cut without appreciably boosting the fuel consumption.
6. Much greater erosion and corrosion resistance (e.g. magnesite–chromite roofs have doubled the campaign of an open-hearth furnace compared with dinas).

The failure mechanism of magnesite–chromite bricks in open hearth roofs is different from that in dinas roofs. Temperature variations make the former material expand and shrink, causing the hot face to shear off (spalling) in layers of 3–5 cm, so open hearth roofs should not be cooled below 1 500°C during service or below 1 100°C during fettling.

TABLE 10.2

*Properties of Typical Magnesite–Chromite Refractories*

| Property | Russian chrome–mag | German chrome–mag | Mag–chrome |
|---|---|---|---|
| MgO | 48·2 | 40·3 | 66 |
| $Cr_2O_3$, % | 22·0 | 25·0 | 11·8 |
| Refractoriness under load, 2 kg/cm$^2$ | 1 510 | 1 560 | 1 510 |
| Spalling resistances 7–15 cycles | | | |
| Apparent porosity | 21% | 26 | 16·2 |
| Compressive strength, kg/cm$^2$ | 575 | 134 | 467 |

Unfired brick is made with and without metal cladding or cases. The clad shapes, usually in the form of large blocks, are used for open-hearth walls and in electric steel furnaces above the slag level. The unclad bricks are used in open-hearth roofs.

## Creep and volatilisation

Recent research has shown that in basic refractories such as chrome–magnesite certain constituents may be volatilised from the surface and the underlying layers. The loss in weight of the refractories may fall off quite markedly in time and in inert gas atmospheres, such as helium, under experimental conditions, the rate of volatilisation is almost the same as in air. The loss of volatile constituents, *e.g.* chromium, from refractories in service may impair or alter their working properties. Borobkov *et al.* (1969) studying the deformation properties of basic refractories, including magnesite–chromite, showed that they have relatively high creep velocities even with small loads and at temperatures of around 1 400–1 450°C. At higher temperatures, in addition to deformation processes, the researchers detected volatilisation of basic refractories due to the presence in them of oxides with a high vapour pressure. This referred in particular to chromite refractories from which the chromium oxide, chrome spinel and other compounds tended to volatilise. Volatilisation depends on the quantity of volatile compounds present and the service conditions such as temperature, gaseous atmosphere, the speed of the gas current, etc.

## REFERENCE

Borobkov, L. B., Lukin, E. S. *et al.* (1969). *Ogneupory* No. 1, 37–43.

*Chapter 11*

# Forsterite

Forsterite refractories are chemically basic materials containing up to 85% forsterite mineral (with a melting point of about 1 900°C) and up to 15% magnesioferrite. They are produced from a range of raw materials which includes olivinites, serpentines, talcs and dunites. Magnesia (magnesite) is added to the batch and the processes occurring during firing are determined by the dehydration and oxidation of the raw materials and their reaction with the added magnesite. The product is a more highly refractory material than the starting components.

The density of forsterite refractories may be increased by certain additives such as $TiO_2$ (for talc-based bodies); $P_2O_5$ in combination with CaO, $Na_2O$, or NiO in order to encourage forsterite formation; $B_2O_3-P_2O_5$ mixes for talc–magnesia–dunite batches and $TiO_2-ZrO_2$ mixtures for siliceous magnesite.

Dunite (typical composition 34·3% $SiO_2$, 44·5% MgO, 6% $Fe_2O_3$, 1·6% FeO, 0·20% $Al_2O_3$, 0·45% CaO and 13% loss on ignition) is commonly used to make forsterite refractories. Talc, containing 64% $SiO_2$, 32% MgO, and iron and alumina impurities, is another forsterite-producing raw material but is not in common use. Serpentine is the main material used in Western Europe, and dunite is commonly employed in the Soviet Union.

## PRODUCTION

Dunites are prefired at 1 400–1 500°C and blended with the magnesite grain, then pressed, dried, and fired by the conventional ceramic process. Olivines may be used without precalcination. Sufficient magnesia must be added to the other raw materials in the batch in order to convert all the silica to forsterite and the iron oxides to magnesioferrite; any excess MgO will tend to improve the refractoriness of the fired brick. Usually 15–25% MgO is added.

As with magnesite–chromite refractories, forsterite can be used for furnace building in the unfired state, since it undergoes only slight volume

117

changes. Slip-cast forsterite has also been made, and this is very dense, of low porosity, and has a high refractoriness under load. It does not shrink during repeated firing.

As in all ceramic refractories (in contrast to those made by non-ceramic methods, such as fusion casting), the role played by the bond is critical in forsterite refractories. Forsterite mineral ($2MgO.SiO_2$) is actually the bond in chrome–magnesite refractories (q.v.). It is also present in highly siliceous magnesite. The properties and behaviour of forsterite as the bond, or as a minor constituent in other refractories, are also of great practical importance to the refractories technologist. In magnesite, for instance, the sudden collapse of the brick upon reaching its point of failure (in contrast with aluminosilicate refractories, which slump over a wide temperature range) is due to the behaviour of the forsterite bond. The forsterite is solid at a certain temperature, then quickly becomes liquid with a slight rise in temperature.

To ensure that the bond does its work and sinters the refractory grains, it is usually incorporated as a finely milled blend of magnesium–silicate component (that is, the forsterite raw material) and magnesia. A green (temporary) bond is added in the form of sulphite lye, and the batch is pressed at 850–1 000 $kg/cm^2$ to yield a green brick having a density of 2·5–2·6 $g/cm^3$. Forsterite brick is more troublesome to press than magnesite brick because of 'spring back'.

Firing is done in tunnel or periodic kilns at about 1 650°C in an oxidising atmosphere. Such high temperatures are needed because, although the forsterite-forming reactions are almost complete at 1 450°C, the processes of sintering and essential crystal growth that confer valuable refractory properties are very sluggish. The crystal syngony (rhombic) of forsterite also hinders sintering.

**Physicochemical processes**

When the products formulated on the basis of the above methods are fired, the olivine, talc or serpentine decompose; all these contain divalent iron which isomorphically replaces magnesium. In oxidising conditions the ferrous oxide is oxidised to the ferric state. The olivine decomposes to give metasilicate and pyroxene (metasilicate with some free silica). There is an increase in volume and a reduction in the basicity of the silicates; the materials are embrittled and the porosity rises. These reactions start at about 500°C and may be completed by 1 400°C, although they are sluggish, and in the firing of commercial products another 250°C is given with a long soak. Above 1 200°C the ferric oxide may change to magnetite.

From the phase diagram for $MgO–SiO_2$ it can be seen that the MgO forms two silicates: forsterite (orthosilicate) with the formula $2MgO.SiO_2$ and metasilicate, $MgO.SiO_2$. The metasilicate is not a refractory compound (its melting point is 1 557°C). However, the forsterite phase is a refractory material. Moreover the metasilicate fuses incongruently at 1 557°C to form the desirable forsterite and free silica:

$$2(MgO.SiO_2) = 2MgO.SiO_2 + SiO_2$$

But below the temperature of 1 557°C the metasilicate combines in the solid phase with the added magnesia:

$$MgO.SiO_2 + MgO = 2MgO.SiO_2 \text{ (at 1 450°C)}$$

to yield more forsterite.

The iron oxides present in the raw materials are changed into the spinel, $MgO.Fe_2O_3$ (magnesioferrite) which has a melting point of 1 780°C and is also a highly desirable refractory.* Various raw materials produce varying amounts of the different phases upon firing. Table 11.1 indicates the

TABLE 11.1

*Phase Compositions of Forsterite Raw Materials Fired at* 1 450°C, %

| Minerals | Olivine | Talc–Magnesite | Dunite |
|---|---|---|---|
| Forsterite | 56 | 54 | 60–75 |
| Clinoenstatite | 19 | 20 | 12–20 |
| Cordierite | 3 | 6 | 1–2 |
| Magnesioferrite | 20 | 18 | 12–18 |

approximate phase compositions of olivine, talc–magnesite and dunite materials now being used in various countries for making forsterite refractories.

In addition to the use of the natural materials mentioned above, it is possible to make forsterite refractories from quartzite and magnesite. For example, a 'synthetic' forsterite can be made from caustic magnesite, very fine silica, and quartz sand. The technique involves briquetting the batch and firing at high temperatures. Talc–magnesite batches are also briquetted and fired hard to give very dense forsterite.

## PROPERTIES AND USES

Properly formulated forsterite refractories are highly refractory and have a high refractoriness under load. They are of interest to the steel manufacturer who uses them for a range of applications, *e.g.* checkers and open-hearth downtakes. They are used also for copper smelting in reverberatory

* $Mg_2SiO_4$ $xFe_2O_3$ = $MgO.SiO_2$ + $MgO.Fe_2O_3$ (starts at 1 200°C and finishes at 1 450°C). The expansion is slight.

furnaces. Forsterite produced in the early period of its development as a commercial refractory *per se* (in contrast to its use as a bond for other refractories: *see* above) was not very highly slag resistant, although fabrication technology has now improved this property.

Having a similar thermal-conductivity value to silica (dinas), forsterite bricks have a much higher wear-resistance than silica when used in the furnace walls and roofs of copper reverberatory furnaces. In fact, some furnace builders now suggest that for this purpose only the hottest part of the roof and around the charging funnels should be built with refractories other than forsterite (*e.g.* magnesite–chromite). Forsterite is cheaper than chrome–magnesite.

The spalling resistance of forsterite is not high, especially when the products are very dense, chiefly owing to the great anisotropy of the thermal expansion across the three axes of the forsterite crystal: $(13\cdot6, 22,$ and $7\cdot6)10^{-6}$. The average $\alpha$ value is about $11\cdot5 \times 10^{-6}$.

The designer of rotary cement kilns is also paying more attention to the advantages of forsterite for kiln linings. Used in the calcination zone, forsterite encourages the development of a protective slag or skin, using the cement batch for its formation and thus prolonging the furnace life (*see* Chapter 21).

Forsterite, being a basic refractory and highly reactive towards acid refractories, should not be laid adjacent to silica and firebrick in linings that are fired above 1 500°C. The reactions taking place would produce highly fusible compounds, leading to early failure of the structure.

Forsterite is also used in conjunction with other materials such as chrome and alumina. Spalling resistant magnesia–forsterite refractories can be made, for example, from 30% raw dunite and 70% magnesite. They have an apparent porosity of 22%, a refractoriness-under-load at 2 kg/cm$^2$ of 1 680–1 700°C, and a spalling resistance of about 20–50 water cycles.

Chrome–forsterite brick also has a much higher spalling resistance than forsterite. Both chrome and magnesia–forsterite products are now being used in open-hearth checkers.

*Chapter 12*

# Dolomite

Dolomite refractories are those containing calcium oxide and magnesium oxide, together with varying amounts of other calcium compounds as impurities. The mineral dolomite contains equimolecular amounts of $CaCO_3$ and $MgCO_3$ but the term 'dolomite' is also given to magnesian limestones.

Thus, dolomite refractories may contain the following compounds: $CaO, MgO, 3CaO . SiO_2, 2CaO . SiO_2, 4CaO . Al_2O_3 . Fe_2O_3, 2CaO . Fe_2O_3$, and $3CaO . Al_2O_3$. All dolomite refractories are chemically basic. The PCE depends mainly on the free lime concentration and periclase content. The refractoriness increases with the concentrations of these oxides. The fusing points and concentrations of melt can be quite accurately read off the phase diagrams.

## DOLOMA

The word 'doloma' is used in Britain to describe the material formed when dolomite is calcined at about 1 700°C in cupola or vertical shaft furnaces, according to the reaction:

$$MgCa(CO_3)_2 \rightarrow CaO + MgO + 2CO_2$$

The decomposition occurs in two stages: with calcium carbonate forming as an intermediate phase with some MgO and $CO_2$.

The word 'doloma' indicates that the material contains the oxides obtained from dolomite just as the word 'magnesia' denotes the oxide obtained from magnesite (that is $MgCO_3$).

Since the calcination of dolomitic raw materials yields free lime, as shown by the above reaction, one problem with their use as refractories is the capacity for hydration and 'perishing' (loss of cohesiveness and strength). Thus, dolomite refractories can be divided into two classes, depending on whether or not the calcined material has been processed to deal with the problem of hydration:

121

(a) Water resistant (no free lime).
(b) Hydratable (with free lime).

The forms in which dolomite refractory materials are used in furnace constructions are: fired and unfired brick, blocks and shapes; unfired bricks, etc.; powders (burnt or raw); and as rammed linings.

When treated with tar, dolomite forms a special class of water-resistant refractories (containing free lime) for use particularly in oxygen steel converters (*see* Chapter 19). The tar coats the grains of burnt dolomite and seals off the access of moisture from the atmosphere.

Impurities in dolomites (apart from those added by the manufacturer to aid mineralisation and stabilisation) include $Al_2O_3$, $Fe_2O_3$, $SiO_2$ and $Mn_3O_4$. Alkalis may also be present: they retard sintering even though producing a liquid phase.

Alumina is particularly harmful and clayey materials in the raw dolomite are usually washed out before use.


## STABILISATION OF DOLOMITE

Mention has already been made of the use of tar to protect burnt dolomite grains against hydration and to reduce the 'perishing' fault. It is also possible to stabilise dolomite by incorporating additives such as phosphates, boric acid, etc., which hinder certain phase changes during calcination tending to produce compounds that are highly susceptible to hydration. Thus, by adding $P_2O_5$ to dolomite, the calcium oxide is mostly converted to $3CaO.SiO_2$ and any dicalcium silicate is stabilised.

Serpentine is used with the same purpose in mind and in this case excess serpentine is incorporated to ensure the bonding of all free lime with the silica contained in the serpentine. This again complicates matters by yielding dicalcium silicate which must be rendered unhydratable with inhibitors such as $P_2O_5$.

When dolomite is burnt a dense skin forms on the grains. Provided the necessary impurities are present the composition of the skin may be such as to enhance the antihydration properties. The phenomenon forms the basis of recommendations and patents for processes to render calcined dolomite resistant to hydration. For instance, Japanese Patent No. 3688 (1954) describes compositions in which a skin is developed on the grains as a result of mixing iron oxide with the raw magnesite before calcination. The skin contains the following compounds: $MgO.Fe_2O_3$, $2MgO.2Al_2O_3$ $.5SiO_2$, $MgO.SiO_2$ and $2MgO.SiO_2$.

In other countries, dolomite deposits with high iron-oxide contents are used for making tarred dolomite refractories so that they will resist hydration in storage and transport.

A third possibility is to use reactive silica in place of forsterite. The silica then bonds directly with the free lime.

Hydration-resistant dolomite refractories produced from materials that have been prepared on the basis of the above calcining technology usually contain about $16 \cdot 5\% \ SiO_2$, $11 \cdot 2\% \ Al_2O_3$, $2 \cdot 7\% \ Fe_2O_3$, $0 \cdot 75\% \ Mn_2O_3$, $46\% \ CaO$, $30\% \ MgO$ and $1 \cdot 2\% \ P_2O_5$.

TABLE 12.1

*Composition of Some Dolomites*

| Country of deposit | Chemical Composition % | | | | | Loss on ignition |
|---|---|---|---|---|---|---|
| | CaO | MgO | $SiO_2$ | $Fe_2O_3$ | $Al_2O_3$ | |
| England | 52·5 | 42 | 1·2–1·5 | 1·8 | 2–2·5 | — |
| USA | 51·7–55 | 38–41 | 0·8–1·5 | 4·25–5·0 | 0·67–0·77 | — |
| Japan | 32–62 | 54–28 | 1·4–3·6 | 2–4·2 | 1–1·2 | 1·6–4·7 |
| West Germany | 56–60 | 38–42 | 0·5–0·8 | 0·7–0·9 | 1·4–1·8 | — |
| Austria | 10–57 | 41–80 | 0·6–5·0 | 0·85–6 | 1–1·6 | 0·5–2 |
| Brazil | 56·7 | 32·7 | 1·8 | 1·2 | 1·7 | 5·9 |

Such refractories last for about 10–12 months in the backwalls of an open-hearth furnace and for about 26–65 heats in the walls of electric steel furnaces. The spalling resistance of such bricks is high and if the powder is used for fettling (repairing furnace linings) it adheres well to the hearth to form a sound monolith.

Table 12.1 shows the chemical composition of some raw dolomite deposits in various countries as used for making tarred refractories.

## PHASE COMPOSITION AND PROPERTIES

The phase composition of calcined dolomite used for brick making and especially for repairing furnaces (fettling) has a decisive effect on the life of the lining. As might be expected, the formation of large quantities of liquid phase should be avoided. However, if no liquid phase at all is formed during calcination of the dolomite grain, the essential sintering process is severely restricted and excessively high temperatures are needed, both for producing the fettling powders and for executing the fettling in steel furnaces.

It is found that sintering depends on the concentration of calcium aluminoferrites when the free lime content is fixed. This explains the effectiveness of the stabilising methods, involving iron-oxide and the other additives mentioned above. To improve sintering and increase hydration resistance then, the dolomite's content of sesquioxides and silica should be carefully controlled.

The high refractoriness of dolomite refractories is due to the development in them, during manufacture, of magnesia, calcia and tricalcium silicate which all fuse above 2 000°C. On the other hand, the fluxing agents which melt below 1 500°C are the solid solutions $4CaO.Al_2O_3.Fe_2O_3$ and $2CaO.Fe_2O_3$. As mentioned above, tricalcium aluminate may also be present.

## FETTLING DOLOMITE FOR STEEL FURNACES

Calcined dolomite powder, crushed, milled and screened to produce an optimum grain-size distribution, is used for making hot repairs to basic open-hearth furnaces and electric steel furnaces. The powders are put into place in the hot furnace using special machines at temperatures of 1 500–1 650°C. Some parts of these furnaces are repaired with mixes of burnt dolomite and magnesite. Suitable dolomites for fettling contain 37–38% MgO, 54–57% CaO, 0·50–1·3% $SiO_2$, and 1–3·25% $R_2O_3$.

Some sections of the open-hearth furnace may be fettled with raw dolomite.

## UNFIRED BRICKS FOR OXYGEN STEEL CONVERTERS

Dolomites that have low silica contents and that are difficult to sinter are used for making unfired tar-bonded dolomite products. Iron scale or slag is used as the sintering agent in the calcination process.

Extra hydration resistance is obtained by burning the materials in rotary kilns with scale and other sintering agents to produce granular sinter. The coal-tar pitch used as the bond (impregnating agent) is carefully specified to meet rigid requirements as regards water content, boiling fractions, pitch and naphthalene concentrations, coke residue, specific gravity and the fusing point of the pitch as a whole.

Large tar-dolomite blocks are made by tamping with pneumatic hammers or on vibrating presses, while monolithic linings in steel converters, for instance, are laid with hammers and hot tamping to improve the structure.

The shaped tar-dolomite products should be stored in refrigerated chambers if possible, and should not be kept for more than 2–4 days, depending on the season of the year, before installation in the converter.

## MAGNESITE–DOLOMITE BRICKS

Various proportions of dolomite and magnesite are used to produce hydration-resistant magnesite–dolomite refractories. The clinker is obtained as for dolomite products; wet mixing and burning in rotary kilns.

Quartz (silica) and phosphoric oxide are added as for dolomite burning. The aim is to convert as much of the calcium as possible to the refractory tricalcium silicate.

Magnesite–dolomite bricks are pressed and fired at 1 650°C for 3 h at maximum temperatures (in tunnel kilns) or at lower temperatures for longer periods (25 h) in periodic kilns.

Unfired magnesite–dolomite refractories are also made, often on site, by mixing ground clinker and pressing or tamping. Since they have hydraulic setting properties, storing after moulding causes a rise in compressive strength from about 100 (after 24 h) to 350 kg/cm$^2$ (after a month). The apparent porosity is 10–12% and the refractoriness under load 1 450–1 600°C (2 kg/cm$^2$).

## Chapter 13

# Heat-insulating Materials

The structure of a refractory material determines its ability to behave as an insulant. It is possible that structure, especially porosity, is more important than chemical composition. All types of refractory compositions can be fabricated to yield insulating structures (Fig. 13.1).

In designing heat insulation the aim is to reduce the thermal conductivity of the material as much as possible (*see* Fig. 13.2). This is usually done by enclosing air in the form of pores in the material to give maximum porosity, consistent with the other properties, *e.g.* mechanical strength and resistance to corrosive gases and slags.

A simple way of making refractories highly porous is to add large quantities of sawdust to the powder batch. When the pressed bricks are fired the sawdust burns away, leaving pores.

FIG. 13.1.  Apparatus for measuring the pore-size distribution of refractories. *Photo:* Courtesy British Ceramic Research Association.

126

Fig. 13.2. Equipment for determining the thermal conductivity of refractory heat-insulating materials. *Photo:* Courtesy British Ceramic Research Association.

Insulating materials are used to save fuel and to enable furnace operators to intensify heating processes and speed up firing cycles.

## RAW MATERIALS

Heat technologists use a wide range of insulating materials. The particular product specified will depend on the maximum temperatures of use and cost. Insulation suitable for hot faces, those in direct contact with the furnace heat, is usually more costly than backing (protective) insulation which does not come into direct contact with flames.

The cost of the insulation depends on the purity and nature of the raw materials and the method of fabrication, since some materials are best shaped and fired by certain methods (*see* below).

The following raw materials are used for insulation:

1. Fireclays and kaolins
2. Silica
3. Basic materials
4. High-alumina materials (*e.g.* sillimanite)
5. Asbestos
6. Vermiculite
7. Diatomite
8. Zirconia, alumina and other oxides
9. Non-oxide compounds, *e.g.* SiC.

## Application temperatures

For convenience, heat insulation can be tentatively divided into the following three classes of service temperature:

(A) *Low temperature* (less than 900°C) — Diatomite, asbestos and vermiculite products

(B) *Medium temperature* (900–1 200°C)

(C) *High temperature* (Above 1 200°C)

All other types of refractories with various degrees of purity and quality

## Castables and concretes

Monolithic linings and furnace sections can be built up by casting refractory insulating concretes, and by tamping into place certain lightweight aggregates suitably bonded. Other applications include the formation of the bases of tunnel kiln cars used in the ceramic industry. The ingredients are similar to those used for making piece refractories, except that concretes contain some kind of cement, either Portland or a high-alumina cement, such as Secar or Ciment-Fondu.

## Methods of fabrication

Naturally occurring massive diatomite (kieselguhr) or diatomaceous earth, can be quarried and used as blocks or powder for making concretes. It has a natural porosity of about 70–80% and can therefore be used without processing, although it is frequently shaped into bricks.

All other materials need to be processed and shaped to give the required porous structures. The two main methods of inducing porosity are:

(a) Combustible additive method (*e.g.* sawdust).
(b) Foaming method.

A third method in which gases are generated *in situ* (chemical method) is also used.

The procedures for making some common insulating refractories are now briefly described, from which the reader may derive an idea of the general principles of the technology.

## KAOLIN AND FIRECLAY INSULATING BRICKS

### (a) Combustible additive method

Various organic materials may be used as the combustible additives, for instance, sawdust, husks of various crops such as rice, cork, cocoa bean remnants, etc. Recently experiments have been done with the use of cotton plant stalks for making insulating refractories. The type of material must burn out readily during the normal ceramic firing process.

Hardwood sawdust such as oak is considered to be better than softwood sawdusts since the physical properties of the former contribute to the refractory shaping processes. The grain-size distribution of the combustible additives is important in determining the pore sizes and the overall porous structure of the fired refractories. Coke of various types and grain-size distributions is also used as the combustible additive. This material is often preferred to sawdust since during pressing it does not cause the powder to 'spring back'.

The porous structure of kaolin and fireclay insulating materials is mainly produced by using combustibles but it also may be helped by incorporating specially made porous grogs and chamotte. Ordinary dense grogs are generally employed and in this case the porosity is developed solely by means of the combustible additive. Grog concentrations are 15–25%.

The properties of the clay in making these relatively non-plastic bodies are very important. Sometimes the clay is added in the form of a slip which envelopes the particles of chamotte and sawdust or other combustible, giving a more homogeneous structure.

The ingredients are blended in runner mills and other equipment, with moisture contents of 20–36%. The resulting mixture is then fabricated by semi-dry pressing or by wet extrusion. The shaped and dried products are then fired in tunnel or periodic kilns in strongly oxidising atmospheres to ensure the removal of all the organic pore-inducing materials.

It can be seen that the method of making fireclay and china-clay insulating products differs only slightly from that used to produce dense refractories. The main difference, of course, is the presence of combustible, pore-forming materials in the insulating bricks. The materials obtained by this method have densities of $0.8–1.5$ g/cm$^3$, depending on the proportions of organic and inorganic constituents. Very light materials with densities of as low as $0.25$ g/cm$^3$ have been made by using cellulose derivatives and paper as the combustible materials.

### (b) Foaming method

The pores required for insulating materials in this case are induced as a result of the use of a foam. In fact the foam is the main constituent of the body. The clay, combustible additive and grog are blended with water to obtain a slip. This is then mixed with the chemically produced foam which has been specially prepared. The foamed mixture is then poured into moulds where it is allowed to set prior to firing (*see* Fig. 13.3). Suitable foaming agents are produced by the chemical industry and when used in conjunction with resins and various adhesives produce very stable foams, giving a range of porosities.

The method of making foamed insulating bricks is illustrated in the accompanying flow sheet. One type of foaming agent consists of rosin oil produced from 45% rosin and 7% sodium hydroxide, the rest being water. The foam is stabilised with common fish glue and alum.

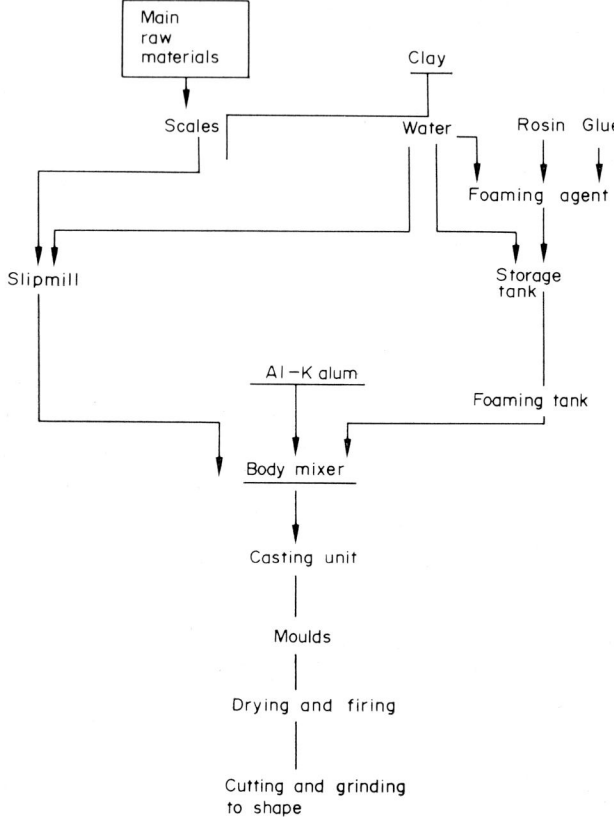

FIG. 13.3.    Flow sheet for foam method of making insulating bricks.

The foamed slip is poured into lubricated metal moulds lined with sheets of paper to facilitate removal. These moulds are then dried on shelf or car driers prior to removal and final drying in the open. Common fireclay and kaolin insulating products made by this method are fired at 1 220–1 380°C. The blocks of foamed material are placed on tunnel kiln cars and after firing are sliced into the required commercial shapes and sizes. Silicon carbide and other cutting wheels are used on cutting conveyers for this job.

Foamed insulating materials are produced with apparent densities of $0.25–0.9$ g/cm$^3$.

*Foamed oxide refractories*
The foam method of making porous refractories can be applied to pure oxides such as alumina and zirconia. These products are suitable for operation at very high temperatures, that is, as hot-face insulation. The foaming agents may be traditional substances such as rosin oil and glues or may involve the use of recently developed foamed polystyrene obtained by emulsion and other methods, using peroxide initiating materials. The service temperature of zirconia insulation may be as high as 2 000°C. The refractoriness under load of alumina insulating products (containing about 98–99% Al$_2$O$_3$) ranges from 1 380°C for an apparent density of about 0.6 g/cm$^3$, to 1 700°C for an apparent density of about 1.0 g/cm$^3$ (with a load of about 0.6–1.0 kg/cm$^2$).

**(c) Chemical method**
In this method the pores are formed by the action of gases generated in the slip as a result of chemical reactions. For example, in the production of common fireclay products the reaction can be made to occur between dolomite and a solution of sulphuric acid according to the following equation:

$$MgCO_3 . CaCO_3 + 2H_2SO_4 = MgSO_4 + CaSO_4 + 2H_2O + CO_2$$

The generation of the carbon dioxide causes the volume of the slip, when cast into a mould, to double. The structure is stabilised by adding gypsum, which when set forms dihydrated gypsum and fixes the pore structure in the mass. The technique is not widely used for making common insulating materials.

## SILICA INSULATION

Silica insulating products have been made on a limited scale by various methods. The earliest technique consisted of adding paper-making materials to ground quartz. The modern method involves the use of coke and anthracite, milled in runner mills and added in amounts of 30–35% of the body. Foaming or frothing methods have also been used based on the use of lime and aluminium powder to generate hydrogen in order to induce the pores.

The technique of making insulating silica is similar to that used for dense silica in that the quartzite is treated with sulphite lye and mineralising agents such as calcium oxide and iron oxide. Drying and firing follow the usual pattern and the green silica products are set in the upper rows of the kiln cars. The mineral composition of such products is 75–85% tridymite, 12–15% cristobalite and 4–8% quartz. The apparent density of insulating silica ranges from 0·95 to 1·25 g/cm$^3$. The after-expansion at 1 450°C is about 0·2%, and the refractoriness (PCE) ranges from 1 675 to 1 710°C. The refractoriness under load (2 kg/cm$^2$) is 1 600–1 640°C. The advantage of silica insulation over aluminosilicate insulation is the higher refractoriness under load and for certain applications the slight increase in volume (about 0·2%) compared with a shrinkage for fireclay refractories. However, the thermal conductivity of silica insulation is usually higher than that of fireclay insulation. Aluminosilicate insulating products customarily have a greater spalling resistance than silica, and special care is needed if silica brick is subjected to sudden heating changes.

## OTHER INSULATING MATERIALS

In addition to those already mentioned, magnesite, high-alumina, chrome–magnesite and most of the other common refractory materials can be processed to give porous insulation. Recently silicon carbide and other non-oxide refractories have been processed by foam methods for making insulation. Another material which has joined the list of insulants is anorthite and by incorporating graphite into clay mixtures it is possible to enhance the spalling resistance of porous fireclay products.

## THEORETICAL AND PRACTICAL ASPECTS OF INSULATION

The retention and transmission of heat by furnace materials is a highly complicated process and the application of theory is not always successful in producing recommendations for furnace and kiln design and service. This is due to the heterogeneities of the materials, the constantly changing temperature and atmospheric conditions and the physical and chemical changes that occur in the refractory linings and also in the furnace charges (metals, ceramics etc.) during processing.

The thermal conductivity values of commercial refractories determined under laboratory conditions can serve only as a guide for practical application. Comparative tests in which the insulating properties of several possible materials are tested for a certain job are more useful, but even this procedure is not foolproof since it is rarely possible to approximate furnace conditions to those available in the laboratory (*see* Fig. 13.4).

The main theoretical principles of thermal conductivity are discussed briefly in Chapter 2. In practice the conductivity ($K$) values are quoted in various countries in different units, for example B.Th.U/ft$^2$/h/°F/in; and kcal/cm$^2$/sec/°C/cm.

As mentioned above, the structure of insulating refractories is critical. The pore-size distribution and the manner in which it was obtained, the ratio of crystalline and vitreous phases present, and the anisotropic nature of the material all exert an effect on the amount of heat transmitted. In general for any given chemical–mineral composition the higher the porosity (the lower the bulk density) then the greater the insulating ability (the lower the thermal conductivity) of the material.

FIG. 13.4. Thermocouples are embedded into ceramic-fibre refractory to determine the temperature gradient through the structure. *Photo:* Courtesy British Ceramic Research Association.

In comparing different chemical compositions (with a fixed porosity) the pattern is less clear, although there is some evidence to suggest that pure oxides are better conductors than complex systems, *e.g.* magnesia is a better conductor than fireclay.

In any consideration of which insulant to use for a given job the furnace designer is concerned with heat losses and capital costs (that is, the purchase and laying of the insulating structures). The job of calculating heat losses is simplified and much of the theory of conductivity ignored, by the use of curves plotting heat loss versus furnace temperatures and wall thicknesses. Such graphs may be found for various materials in textbooks dealing with heat engineering. Useful data can also be obtained from refractory suppliers.

## Chapter 14

## Fusion-cast Refractories

In manufacturing many types of refractory materials the aim is to maximise their density. The conventional (ceramic) process described in Chapter 4, involves firing granular mixtures. If the firing temperatures are raised to the point where the whole mass fuses into a glass and this melt is then poured into suitably shaped moulds as in metal casting, the density of the products should be truly maximised. These products are known as fusion-cast refractories. Once the melts have been poured into the moulds, crystallisation occurs. The cooled shapes have to be annealed just as glass is annealed to remove stresses that might lead to cracking.

Since the melting of refractory materials in this way is a very costly procedure, it is economical to fusion-cast only highly refractory and relatively pure materials. The process is restricted to alumina, mullite, mullite–zircon, quartz and certain basic materials such as forsterite, magnesium–silicate spinel and magnesite–chromite. The commonest fusion-cast materials are alumina, mullite and zircon compositions.

Fused alumina and mullite are also produced in granular form for subsequent processing by ceramic fabrication. These grains are suitably bonded, either with clay or polymer bonds, and used for furnace construction in the usual way. They are sometimes called fused grain refractories to distinguish them from fusion-cast materials.

Fusion-cast technology was instituted on the basis of alumina but later it was discovered that mixtures of zirconia, alumina and silica (zircon–mullite) had a better resistance to molten glass than alumina alone. Since fusion-cast refractories find their most important applications in the construction of glass-melting furnaces, this discovery was of some significance, and it led quickly to extensive research into the relevant phase diagrams and to important practical developments.

### THE FUSING PROCESS

The electric-arc furnace in which the materials are most commonly fused must deliver a highly fluid melt to the moulds which are usually quite simple in shape. It is impracticable to use mechanical stirring or to rely on convection currents within the fusion itself as a means of homogenising it, so a homogeneous fusion has to be produced by proper choice of the raw materials.

Although electric arc furnaces are most commonly employed in fusion production, it is possible to use electric-resistance furnaces or the heat generated by exothermic reactions *e.g.* by using aluminium powder. Fused quartz ('silica-glass' is a misnomer, since silica has a clearly crystalline structure whereas glass has not) is sometimes melted in resistance furnaces. The exothermic process in which the aluminium powder is ignited to produce very high local temperatures is reserved for fusing alumina.

The batch materials, premixed in the specified proportions and ground finely followed by briquetting, are charged into the electric-arc furnace so that they surround the electrodes. Lumps of coke located between the electrodes act as centres on which the batch begins to fuse. When the fusion processes are to begin, the operator moves one of the electrodes so that it touches the coke and power is supplied to the electrode. The electric arc thus formed is controlled by arranging the pieces of coke appropriately.

The batch materials in the vicinity of the arc created between electrode and coke begin to fuse until a small bath is formed and then melting proceeds steadily by electrical resistance processes within the melt until the whole of the batch is fused. Pieces of unburnt coke are removed in the early stages of the process.

The process of melting highly refractory minerals in this way is difficult and dangerous, and special care and skills are needed. Iron impurities form at the bottom of the furnace (as ferrosilicon) and since these rapidly corrode the lining it is essential to remove them frequently during fusing. The slag is tapped off. The temperatures used for fusing mullite–zircon compositions are 1 900–2 000°C; and for mullite 2 100–2 200°C. The voltages on the electrodes are 120–190 V.

## THE CASTING PROCESS

The aim of casting is to produce a homogeneous block of refractory material, free of cracks, cavities, pores and other features that might reduce its resistance to molten glass and other corrosive media in the furnace in which it is to be used as a construction material.

To avoid the need for expensive machining of cold blocks of fusion-cast materials, mould sizes are carefully specified and the casting and cooling cycles are chosen to reduce shrinkage to a minimum. Moulds made of clay-bonded silica sand (special synthetic compositions and linseed oils are also used as bonds) are filled with the melt at very high temperatures.

Some of the melt is left in the furnace to provide a bath for subsequent batches.

The casting of refractories by this process is more complicated than the casting of metals, mainly because of the much higher temperatures of the refractory fusion at the moment of casting, and the faster cooling once the material is in the mould. Coupled with rapid setting and the high shrinkage (about 10–15% greater than that of most metals) this makes the process skilled but, even so, sometimes unpredictable. Shrinkage cavities in the cast blocks are often very large. They form in the centre of the blocks and may cause sudden failure of the linings in furnaces. Sometimes the cavities are cut out of the blocks which then have to be ground and finished before insertion in furnace linings. Topping up procedures are also used to ensure that a supply of liquid fusion reaches the liquid centre of cooling blocks in the mould. This largely eliminates hollow centres.

The casting of a refractory block of about $0.075$ m$^3$ in volume takes about 2·5–3 min. Smaller parts such as burner port blocks take 30–50 sec. In one version of the process the molten refractory is cast into long slabs of standard brick-size sections which can then be cut, using diamond cutting saws, into the specified brick lengths. Because of the particular technique used the cast slabs are full of holes and these have to be cut out of the material prior to use for furnace construction.

## CRYSTALLISATION, COOLING AND ANNEALING

When the molten charge has been poured from the furnace into the mould the casting has to be cooled. This must be done under controlled conditions; the process can be divided tentatively into two stages:

(a) Crystallisation.
(b) Annealing.

The theory of the crystallisation of glass and ceramic melts is complicated. Much remains to be learnt about what goes on inside a cooling block of fusion-cast refractory. However, the practice of glass annealing is fairly well understood and it has been applied to the annealing of fusion-castings in order to eliminate stresses and the danger of cracking.

Marked variations may occur in the proportions of the various minerals and glass phases remaining in the cooled casting, depending on the chemical composition and the cooling cycle. The variations between different castings may be very great, for instance 5–80% mullite, 5–50% corundum, 5–50% glass, and 5–12% iron and titanium minerals (e.g. magnetite and rutile). The crystal sizes and the degree of homogeneity of the casting may also vary critically in relation to the working properties, depending on the cooling and crystallising rate.

In the production of high-quality mullite blocks it is essential to arrange for a high mullite concentration to develop since if alumina (corundum) is allowed to form it leaves an undesirably high content of glass phase.

Cast and crystallised blocks are usually annealed in tunnel lehrs (in the same way as ordinary glass products) from temperatures of 900–1 000°C to 250°C in about two days. Improper annealing causes faults in the form of cracking, chipped and spalled corners and edges.

The annealed blocks are usually machined (ground) to their final sizes and shape ready for use in furnace linings. Plastic-bonded silicon–carbide abrasives are used for grinding.

## FUSION-CAST MULLITE

The composition range for fusion-cast mullite products is 70–76% $Al_2O_3$, 21–20% $SiO_2$, 1·9–1% $Fe_2O_3$, 0·7–0·25% MgO, 1·5–0·1% CaO, and about 3·5% $TiO_2$. The properties are such that these materials can be used to withstand the corrosive action of a range of molten glasses at very high temperatures. The density is about 3·0–3·3 $g/cm^3$. The compressive strength may be as high as 6 000 $kg/cm^2$ and the apparent porosity is invariably less than 8%. The refractoriness is around 1 820°C and the refractoriness-under-load about 1 720°C.

The spalling resistance of fused mullite is very high. The material will resist acid and basic slags and glasses of a wide range of chemical composition and corrosiveness.

The resistance of fused mullite depends on the presence of other phases in the massive refractory. It is known that clay-bonded granular mullite is much less resistant than fusion-cast mullite blocks because of the presence of the glass and other phases in addition to the main mullite phase. The microstructure of the material is important in other ways. Various structures are known, including the glassy-prismatic mullite structure and the fibrous mullite structure, as well as the granular form already discussed. The fibrous mullite structure appears to be more resistant to melt attack in service. The fibrous mullite crystals form a tangled mass that resists the entry of melt.

By incorporating about 6–7% zirconium dioxide in the basic mullite mixture it is possible to improve the properties of fusion-cast mullite refractories. The zirconium dioxide combines with the silica to produce the mineral baddeleyite which penetrates the glass and therefore increases the resistance of the structure to corrosion and makes it more uniform.

## FUSED ALUMINA

Bauxite and other aluminous raw materials are fused in a manner similar to that described for other fusion-cast refractories. It is also possible to use the thermite reaction mentioned above. The main aim in producing fused alumina refractories is to maximise the density and recent developments have led to the production of fused alumina components with almost theoretical density, that is, completely free of pores. In the practical production of commercial fused alumina refractories for furnace and kiln construction the impurities present cause a glass phase to form, the aim being to produce the maximum quantity of mullite from the silica and alumina in this glass phase, thus increasing the resistance of the alumina refractory. This is achieved by using appropriate cooling cycles in critical temperature ranges.

After being cast into graphite or metal moulds the materials are packed in powdered insulants so that they can be slowly cooled over a period of about 10 days. The apparent porosity obtained with a well conducted process may be as low as 1 %. The refractoriness is about 1 950°C, and the refractoriness-under-load around 1 700°C. When used in service these refractories (heated to about 1 650°C) may shrink by about 0·1–0·5 %, depending on the heating time.


## FUSED SILICA REFRACTORIES


High-quality quartz materials are used for making fused silica refractories; the iron oxide content in particular must be minimal. The quartz is fused in electric-resistance furnaces at a temperature of around 2 000°C to produce cylindrical blocks of opaque silica 'glass'. These blocks, partly solidified, are then removed to moulds where they are pressed at high pressures to make the final product.

The main problems in producing these refractories are to obtain a fully homogeneous structure and to avoid the crystallisation of certain unsuitable forms of silica. In use and during experimental heating, fused silica develops $\beta$-cristobalite at temperatures above 1 100°C. When the temperature rises the rate of crystallisation tends to diminish. Cristobalite formation weakens the block and increases the open porosity. The aim in formulating the compositions and carrying out the technology, therefore, is to develop as much tridymite as possible and to reduce the cristobalite formation. Certain mineralisers are added to achieve this purpose.

The advantages of fused silica (also known as vitreous silica) include very high spalling resistance but this is offset by its ready deformation under load at temperatures of about 850°C. Coupled with the propensity for cristobalite formation at temperatures above 1 100°C, this makes fused silica products of interest only for certain special applications, such as the manufacture of vitreous silica tubes and pipes and reaction vessels used in the chemical engineering industry. Applications calling for a

TABLE 14.1

*Properties of Fusion-Cast Refractories*

|  | Mullite (1) | Alumina (1) | Alumina (2) | Mullite (2) | Silica |
|---|---|---|---|---|---|
| $Al_2O_3$ | 78·4 | 93–96 | 99·3 | 70–75 | 0·1 |
| $SiO_2$ | 20·9 | — | 0·10 | 21–19 | 99·0 |
| $Fe_2O_3$ | 0·07 | — | 0·05 | 0·2–0·7 | — |
| App. porosity, % | 0·5 | 1·1 | — | <10 | 0·0 |
| PCE, °C | 1 900 | 1 960 | 1 900 | 1 730–1 800 | 1 650 |
| Bulk density | 3·11 | — | 3·24 | 2·9–3·3 | — |

combination of high thermal-shock resistance and chemical resistance (to certain reagents) might be met by the use of vitreous silica.

Although vitreous silica is, of course, an acid material and might be expected to react readily with basic compositions, in practice it is found that, provided the temperatures are below 1 700°C and the atmospheres are not reducing, vitreous silica, like other silica materials (ceramic fabricated dinas) actually resists iron-slag corrosion.

## Chapter 15

# Carbon and Graphite

The fact that carbon does not melt, is not wetted by many high-temperature fusions and has a high spalling resistance, would suggest that it is an ideal refractory material. And indeed it is: for certain conditions. The most serious disadvantage of carbon is that it burns away when heated in air or oxygen. Nevertheless, since many heat processes are conducted in reducing conditions, carbon finds extensive use as a refractory.

Some classifications include silicon carbide with carbon refractories. In this book silicon carbide is discussed under carbides in a separate section. However, it is worth pointing out that besides carbon and its pure compounds with silicon, boron, titanium, etc. (all of which are refractories) there are several refractory mixtures which are used extensively in furnace construction. These contain proportions of graphite and fireclay, silicon carbide and fireclay, and chamottes. In addition there are tar-bonded refractories in which organics such as coal-tar pitch are used to impregnate magnesite and dolomite grains, followed by heat treatment in order to coke the tar.

Thus, it is seen that carbon has an important place in refractories technology both as a material *per se*, and as an auxiliary material or additive.

## RAW MATERIALS

Bulk carbon refractories for use in blast-furnace linings and furnaces employed for smelting non-ferrous metals are made from coal, petroleum coke, anthracite, coal-tar pitch, and graphite. Heat-processed anthracite is obtained by subjecting anthracite coals to temperatures of 1 250–1 300°C for periods of 10–20 h in shaft furnaces, or for shorter periods in electric retorts. In shaft furnaces heat is produced from the combustion of some of the anthracite. During this heat treatment, the anthracite is densified and loses most of its volatile constituents. The quality of the anthracite is important in the manufacture of carbon, especially graphite, refractories.

The sulphur content should be less than 2·5% and the ash content less than 4–5%. After calcination the anthracite lumps are cooled in air and classified in sizes for mixing with coal-tar and possibly other additives.

Coke is used in a variety of forms, including petroleum, coal-tar and natural pitch. Preference is given to those forms with low ash concentrations since fuel ash consists mainly of slagging agents (alkaline silicates). Petroleum coke is obtained by heating petroleum residue or heavy crude oil at temperatures of 500–750°C. The coke is obtained in the pyrolysis process (at normal pressures) or in the cracking process (at high pressures).

Coal-tar cokes are obtained by heating mixtures of coals in reducing conditions. Foundry cokes are considered to be the best for refractories production since they are mechanically stronger than blast furnace coke and contain less harmful ash.

Graphite is a pure crystalline allotrope of carbon (the other natural forms of carbon are coal and diamond). Since all three substances consist of the chemical element carbon, it is apparent that the structural differences must be the cause of the markedly different properties.

Graphite refractories are made by graphitising carbon of the coal or common coke variety. Two types of graphitic refractories are produced:

(a) crystalline or silver graphite, very similar to naturally occurring graphite;
(b) cryptocrystalline graphite.

In the Acheson process, used for example by the Union Carbide Company (USA), temperatures of 3 000°C are used to turn anthracite-coal blocks (bonded with pitch) into graphite. Temperatures are controlled in accordance with the electric power supply from the furnace transformers. The production cycle, using carbon electrodes and packing the carbon refractories with coke to guard against oxidation, lasts almost three weeks, mainly because of the long cooling period (two weeks). Prolonged cooling in which the temperatures are very slowly reduced enhances graphite crystallisation.

**Binders and bonding materials**
These consist of coal-tar pitch and tars which after heating produce a coke residue which sticks the carbon grains together to make a strong, monolithic refractory. Bonds used in the production of carbon refractories are evaluated with respect to volatiles, coke and free carbon concentrations.

Coal-tar pitch is a black viscous liquid with a specific gravity of 1–1·25 with highly variable compositions and properties. High quality pitches have less than 2% water and more than 9% free carbon.

Pitch is a shiny solid obtained after distilling the high-temperature fractions of coal-tar pitch at about 350°C. It has a very complicated chemical composition and various types are available, including soft, medium and electrode quality. Pitches soften at 45–105°C.

In some countries wide use is made of artificially prepared coal-tar pitches consisting of about 73·5% coal-tar pitch, 18% anthracene oil, and 8·5% petroleum bitumen. The anthracene oil which dissolves the pitch is a green liquid with a specific gravity of about 1·13, obtained by distilling coal-tar pitch. The mixtures are used to control the physical properties of the pitches used in different refractory mixtures.

## PRODUCTION PROCESSES

Carbon refractories are shaped by extrusion and pressing. After the preparation of the carbon grain by heating, crushing and classification, it is mixed with the binder and fabricated. In extrusion, the carbon-pitch mixes have to be maintained at temperatures at which the binder is in the most suitable plastic state for shaping.

Carbon blocks are made from mixtures of 75–85% carbon (the non-plastic) in the form of coke, and 15–25% binder in the form of coal-tar or a mixture of pitch and anthracene oil (*see* above). The ingredients are blended in blade mixers which can be kept at 100–120°C, using steam heaters. Hydraulic presses (with pressing forces of 100–9 000 tons) are used to shape the mixture.

Carbon bodies suffer from high elastic 'spring-back' and upon ejection from the mould may expand up to 2·5%. The temperatures of the body (90–100°C) during pressing are critical in yielding a dense, uniform product. The pressings are cooled in water as are the extruded sections, prior to firing or impregnation. These blocks are fired in various types of periodic furnace at temperatures of up to 1 500°C. Graphitisation is carried out at temperatures of up to 3 000°C (*see* above: Acheson process). The set blocks are packed with coke filler to protect them against oxidation and combustion.

## PROPERTIES OF CARBON REFRACTORIES

Carbon is a highly spalling-resistant material with a high thermal conductivity. It burns in oxidising atmospheres but has no melting or fusing temperature in the way that silica or clay have. It is volume-stable up to its combustion point and exhibits a good resistance to molten metals, slags, glasses and other fusions. The mechanical strength (refractoriness-under-load) increases with rise in temperature. The following properties have been quoted for blast-furnace carbon blocks:

| | |
|---|---|
| Apparent density | $1{\cdot}55{-}1{\cdot}65$ g/cm$^3$ |
| Apparent porosity | $15{-}17\%$ |
| Compressive strength | $250{-}300$ kg/cm$^2$ |
| After-contraction, $\%$, at: | |
| 1 400°C | $0{\cdot}0$ |
| 1 500°C | $0{\cdot}1$ |
| 1 600°C | $5{-}8$ |
| Coefficient thermal expansion (0–700°C) | $3{\cdot}7 \times 10^{-6}$ |
| Thermal conductivity, kcal/m h deg | $5{\cdot}8{-}6{\cdot}0$ |

Carbon blocks will withstand up to 30 cycles of thermal shock and lose weight only as a result of combustion of the carbon. It is difficult to test this property in laboratory conditions. In contact with iron slags above temperatures of 1 100°C carbon reacts to yield metallic iron and manganese:

$$FeO + C = Fe + CO$$
$$MnO + C = Mn + CO$$

Carbon refractories also react readily with superheated steam (above 500°C), thus:

$$C + 2H_2O = CO_2 + 2H_2 \ (400{-}600°C)$$
$$C + H_2O = CO + H_2 \ (600{-}1\ 000°C)$$

## USES OF CARBON REFRACTORIES

Large carbon blocks weighing up to 750 kg are used in the hearths and wells of open-hearth furnaces (*see* Chapter 19), where their excellent resistance to molten iron is valuable. The producer of non-ferrous metals such as aluminium, antimony, and phosphorus also uses carbon as a cupola and pot lining material. The refractories industry itself employs carbon in furnace design for the production of silicon carbide. Carbon is also used in producing ferrosilicon and other ferro-alloys.

The production of electrodes for high temperature furnaces is an important use of carbon. In aluminium smelting, electrolysers containing molten bauxite are lined with carbon blocks. Furnace builders use very large blocks to reduce joint lengths which are filled with carbon pastes. Carbon is used extensively in chemical engineering.

## CARBON–CHAMOTTE (FIRECLAY) BODIES

Fireclay refractories can be improved by adding graphite which resists slag wetting and boosts spalling resistance. Fireclay–graphite (chamotte–graphite) mixes are used for crucibles, for gutters and transfer channels handling molten iron; for plugs in steel-casting plant and especially in continuous steel-casting equipment.

When added to fireclay, graphite increases its spalling resistance and reduces its thermal expansion factors. In reducing conditions such mixes have a greater resistance to slagging than pure clay brick. A very wide range of compositions is used. High-quality, highly plastic clays are used which vitrify at about 1 200°C. The bodies can be tamped, pressed or extruded from pugmills. Drying is difficult and takes a long time because moisture does not readily leave graphite–clay mixtures.

Firing is done very carefully in coke fillings in saggars or in sealed muffles at temperatures of 1 300–1 360°C.

Such products have the following typical properties:

| | |
|---|---|
| Refractoriness (PCE) | Up to 1 900°C |
| Refractoriness-under-load, 2kg/cm$^2$ | 1 350–1 480°C |
| 10% sag | 1 500–1 600°C |
| Thermal linear expansion factor 0–1 000°C | $(2 \cdot 8 – 3 \cdot 8) \times 10^{-6}$ |
| Compressive strength | 300–500 kg/cm$^2$ |

## CARBON IMPREGNATED REFRACTORIES

The properties of many refractories can be improved for certain jobs by impregnating them with carbonaceous substances such as coal-tar or softened pitch, followed by heat treatment. Another version of the same process involves firing the refractories in currents of gases laden with carbon monoxide and hydrocarbon. It is rather interesting that a process that has long been known to occur, with very adverse effects, in the linings of blast furnaces, is now being beneficially used. I refer to the deposition of carbon particles from carbon monoxide in blast furnace gases.

In the impregnating process the refactories, for example firebrick or magnesite blocks, are placed into autoclaves after being preheated to 200°C, and then the autoclaves are evacuated. Coal-tar pitch at pressures up to 10 atm is then admitted, using pitch with softening points of 110–136°C, and containing about 1·5% water. The impregnation lasts about 15–30 min and the aim is to fill all the pores of the refractory with tar or its decomposition products. No air is allowed into the chamber and so the carbon in the pores is coked. These tar-bonded refractories are used in the steel industry for oxygen converters.

The carbonisation of firebrick is carried out in gases containing hydro-carbons and carbon monoxide at 1 140–1 180°C for 15–25 h so that graphite is deposited on the walls of the pores and amorphous, 'sooty' carbon in the pore spaces. An ancilliary effect of this treatment is thought to be the removal of about 2% of the iron from the firebrick in the form of iron carbonyl. Carbonisation boosts the refractoriness-under-load values of the products.

Graphite–fireclay moulds have been used for casting bimetallic ingots and other special steels. This economises on cast iron and increases the ingot mould life, but, far more important, yields a mirror-finish in the casting so that no further machining is needed.

Graphite–clay (plumbago) refractories are likely to find extensive uses in foundry practice for cores and moulds, metal gutters, for magnetic pumps, nozzles and in centrifugal casting.

## BIBLIOGRAPHY

Blackman, L. C. F. (1970). 'Modern Aspects of Graphite Technology'. Academic Press, New York.
Reinhart, F. (1968). Carbon bodies, *Glas-Email-Keramo-Tech.* No. 12, 425–9.
Walker, P. L. (1966). 'Chemistry and Physics of Carbon'. New York.

## Chapter 16

# Oxide Refractories

Much of the literature on high-temperature oxides is more properly classified under 'Technical Ceramics' rather than 'Oxide Refractories', since the properties of many oxides, such as hardness, chemical resistance, electrical resistivity and the ability to retard neutrons in nuclear engineering applications make them of greater interest to the technical or engineering ceramics user than to the furnace designer. Usually cost and raw material availability are the central factors from the refractory user's viewpoint.

The commonest oxides used by the producers of 'bulk' refractories (excluding those making high-cost, special duty components from such rare oxides as thorium, uranium, tungsten and vanadium, etc.) are alumina, magnesia, zirconia, calcia, and silica.

Beryllia, it is true, is used in large quantities in nuclear engineering because of its low neutron cross section, that is, its capacity to retard slow neutrons without itself absorbing them.

High-temperature oxides exhibit a range of very valuable properties many of which are common to the majority of the compounds in this class of refractories. However, each oxide has distinctive features which determine its range of applications and the methods of preparation for service, including fabrication technology.

## ALUMINA REFRACTORIES

The properties and uses of high-alumina refractories are described in Chapter 7; this section is confined to a brief discussion of pure alumina. Ceramic refractory materials which consist of aluminium oxide with very small traces of impurities are often known as corundum ceramics or pure-alumina. The pure crystal form of this oxide is $\alpha\text{-}Al_2O_3$. The degree of refinement of the alumina-yielding raw materials determines the content of $Al_2O_3$ in the finished product and since the type and amounts of impurities have a critical effect on the properties of the refractory it is

146

essential to specify carefully the total impurity contents of refractories sold as 'pure alumina'.

**Properties**

Alumina is one of the most chemically stable oxides known. It is mechanically very strong, insoluble in water and superheated steam and (when calcined) in most inorganic acids and alkalis. It resists hydrofluoric acid and hydrogen fluoride but at temperatures above 1 700°C it reacts with fluorine gas. Such properties make it suitable for the shaping of crucibles for fusing sodium carbonate, sodium hydroxide and sodium peroxide. It has a high resistance in oxidising and reducing atmospheres even at temperatures up to about 1 850°C (in oxidising conditions it melts at 2 040°C).

Alumina can be decomposed by fusing it with potassium bisulphate but when fused it can be dissolved only in sulphuric acid under pressure. Its resistance to molten metals at temperatures up to 1 000°C, such as bismuth, phosphorus, arsenic, and also sulphur and its compounds, makes it useful in high-gravity metallurgical processes.

Alumina is used to make thermocouple sheaths for high-temperature measurements in molten metals and slags, where its gas impermeability at temperatures up to 1 720°C is very important.

*Compressive strength* of alumina at room temperature is very high (of the order of 30 000 kg/cm$^2$), but this property falls markedly as the temperature rises: to about 8 000 kg/cm$^2$ at 1 000°C, and down to about 500 kg/cm$^2$ at 1 600°C.

*Tensile strength* values at the above mentioned temperatures are much lower, respectively: 1 400, 650, and 350 kg/cm$^2$ (at 1 200°C for the last quoted figure).

The properties of alumina, like those of most refractories, depend very much on the texture and particularly the porosity of the fabricated article. Alumina technology has been perfected to such a degree that structures ranging from almost the theoretical density of corundum down to a porosity of 90% (for heat insulation purposes) can be obtained. The incorporation of additions such as magnesia is common in developing certain types of refractories based on alumina.

*The thermal-shock (spalling) resistance* of pure alumina refractories of high density is only moderately good. This property is closely related to the nature of the crystalline structure of the refractory and in coarsely crystalline articles it is 2–4 times higher than in finely crystalline. The *thermal expansion* of sintered pure corundum in the range up to 1 000°C is 8–8·5 × 10$^{-6}$. There are no high-temperature polymorphic inversions and so as the temperature rises the expansion can occur uniformly.

The thermal conductivity (on which spalling resistance partly depends) appreciably diminishes with a rise in temperature, a feature of most crystalline materials. The conductivity of dense alumina refractories may be cut to a fifth with a rise in temperature to 1 000°C.

## Methods of fabrication

Pure alumina refractories are shaped by pressing or casting. Recently isostatic pressing has also been used. In order to reduce the firing temperature of fabricated alumina refractories it is possible to make additions such as 2–4% by weight of a mixture of $Mn_2O + TiO_2$ or $Cu + TiO_2$. This yields products with a bulk density of about 3·8 kg/cm$^3$ when the firing temperature is less than 1 400°C.

In order to obtain products with very high densities (bulk density up to 4·0 kg/cm$^3$) hot pressing using graphite moulds, pressures of about 400 kg/cm$^2$ and temperatures above 1 700°C are used. At the other end of the scale, foaming methods of fabrication will yield heat-insulating alumina materials with a true porosity up to 80%.

## Uses

The above list of outstanding properties of alumina indicates that the material finds extensive use in most heat-processing industries. Highly porous alumina insulating materials can be used for lining gas and electrical kilns and furnaces operating at temperatures up to 1 850°C. Muffles and protective pipes and tubing with various surface shapes can also be made from alumina for use as winding cores taking tungsten and molybdenum wire for use in electrical furnaces operating in hydrogen or inert gases. The use of alumina for the fabrication of a wide range of crucibles is well known: these are used for melting metals, glasses, oxides and slags. Other uses include the manufacture of thermocouple sheaths, laboratory ware and chemical apparatus. Bulk alumina refractories made by electrofusion casting techniques are used in the glass industry.

# BERYLLIA REFRACTORIES

The use of beryllia oxide as a bulk refractory material is limited because of its proneness to volatilisation, especially in the presence of water vapours. Such volatilisation may become quite marked even at temperatures as low as 1 000°C. Beryllium oxide has a melting point of 2 570°C, a high mechanical strength and a very high thermal conductivity. The mean linear coefficient of thermal expansion between 25 and 1 400°C is about $9·3 \times 10^{-6}$ and in the range up to 1 700°C, $10·6 \times 10^{-6}$. Up to 2 000°C the expansion factor is $10·1 \times 10^{-6}$.

Chemically, beryllium oxide is inert to hydrogen peroxide, nitrogen, carbon dioxide, carbon monoxide, sulphur dioxide, hydrogen, bromine, iodine and ammonia. It readily reacts with fluorine, however, to form beryllium fluoride and there is a slight reaction with chlorine at room temperature. Beryllium oxide dissolves readily in hydrochloric and nitric acids, but with sulphuric acid its reaction is weak. It rapidly dissolves in hydrofluoric acid. Beryllium oxide crucibles are used for melting phosphate slags and similar glasses. Molten lead and molten borates have little effect on beryllium oxide but in contact with acid substances it is unstable and beryllium oxide crucibles are unsuitable for melting glasses commercially. Beryllium oxide is readily reduced to the metal by other metals having a marked affinity for oxygen, such as magnesium and calcium.

The thermal-shock resistance of beryllium oxide refractories is higher that that of articles made from most other highly refractory oxides. This is mainly because of its excellent thermal conductivity.

*Fabrication methods* for beryllium oxide refractories include pressing, sintering and machining. Before pressing special bonds, including water-soluble synthetic resins, are impregnated into the powder. Fabrication pressures are as high as $3\,500\ kg/cm^2$, with isostatic pressing up to $10\,000\ kg/cm^2$. Beryllia can also be cast from water slips at water concentrations of up to $70\%$ by weight and at pH values of 1–2. Porous insulating products can be made by the combustible-additive method or by chemical foaming techniques. A common combustible material is petroleum coke.

Commercial refractories made from beryllia can be cut with diamond saws using water lubrication. The sintered refractories can also be machined on grinders using abrasive wheels. Relatively low-density beryllia items can also be machined on the lathe using high-speed steel tools.

*Uses of beryllium refractories* include the manufacture of a wide range of crucibles for melting uranium and thorium. As previously mentioned this material finds extensive use in nuclear engineering.

## CALCIUM OXIDE REFRACTORIES

All calcium oxide articles must be made without allowing the material to come into contact with water because of the capacity of this material for hydration. Over the years many attempts have been made to reduce the hydration tendencies and to stabilise it. Various materials are added to calcium oxide before, during and after calcination, including titanium dioxide, zinc oxide, zirconium dioxide, thorium dioxide, and tin oxide, although most of these have little influence on its hydration properties. Certain materials, including iron oxide, chromium oxide, cobalt oxide and nickel oxide, tend to increase the hydration resistance but possibly the best technique is the impregnation or surface treatment of the fabricated calcium oxide crucibles and shapes in order to prevent hydration.

**Properties**

The chemical properties of calcium oxide are discussed in detail in many general chemistry text books and there is no need to dwell on this subject here. For the refractories user, it is relevant to note that the melting point of CaO is 2 600°C. (Some researchers put it at 2 620°C.) At high temperatures calcium forms calcium carbide, $CaC_2$, with carbon and in the system $CaO–CaC_2$ there are several chemical compounds with melting points around 1 980°C and two eutectics containing 69% and 35% of $CaC_2$ with melting points, respectively, of 1 750 and 1 800°C.

**Uses of calcium oxide refractories**

Stabilised calcium oxide products can be used for lining rotary furnaces for smelting phosphate ores. Crucibles are widely used for smelting high-purity metals and rare metals such as platinum.

## MAGNESIA REFRACTORIES

Magnesium oxide melts at 2 682°C. It has a high volatilisation resistance; even in oxidising and neutral atmospheres volatilisation is not noticeable up to a temperature of 2 000°C. However, in the presence of carbon dioxide or carbon, volatilisation increases appreciably at 1 800°C.

**Uses**

Magnesium oxide is used for crucible fabrication for melting a range of metals including beryllium, aluminium, uranium, etc. Very high-purity manganese can be obtained in these crucibles. Magnesium oxide refractories are used in bulk for lining high-temperature furnaces, particularly for the combustion of atmospheric nitrogen at temperatures up to 2 200°C. Large single crystals of magnesia can also be employed for making certain critical components of guided missiles.

## ZIRCONIA REFRACTORIES

Information of interest to the user of zirconia refractories is given in Chapter 14, which describes fusion-cast refractories. Since zirconium dioxide, $ZrO_2$, is a polymorphic material, there are certain difficulties with its use and fabrication as a refractory material. It is essential to stabilise it before application as a refractory. This is achieved by incorporating small additions of calcium, magnesium, yttrium and cerium oxides. If raw zirconium dioxide were to be used without stabilisation, during heating and cooling the crystalline inversions taking place in the material would cause disruption of the structure.

**Properties**

These depend mainly on the degree of stabilisation and the type and quantity of stabiliser as well as on the quality of the original raw material. Sintered zirconia refractories have a very high strength at room temperature which is maintained up to temperatures as high as 1 500°C. They are, therefore, very useful as high-temperature constructional materials for furnaces and kilns. The thermal conductivity of zirconium dioxide is found to be much lower than that of most other refractory oxides, and the material is therefore used as a high-temperature insulating refractory.

Since zirconia exhibits very low thermal losses and does not react readily with liquid metals it is particularly suitable for making refractory crucibles and other vessels for metallurgical purposes. It shows a negligible reaction with molten steel, finding uses in the steel industry as well as in the smelting of such metals as palladium, ruthenium and rhodium. As mentioned above, zirconia is a particularly valuable refractory material for the glass furnace designer, mainly because it is not easily wetted by molten glasses and because of its low reaction with them. Zirconia has a low thermal conductivity and is, therefore, useful for making insulating refractories for furnace linings operating above 2 000°C. Zirconium dioxide and its solid solutions are sometimes used for making electrical heating elements and for building induction furnaces. The use of zirconia as a bulk refractory for making massive furnace linings is restricted by the cost of the material; it is more customary to use the cheaper zircon (zirconium silicate). The properties and uses of zirconium silicate are discussed in Chapter 14.

## SILICA REFRACTORIES

Silica (or dinas) refractories are extensively used for furnace and kiln construction. Their properties and uses were described in Chapter 8.

Fused or vitreous silica, sometimes known as siliceous or quartz glass, is an important refractory material for certain special applications. It melts at 1 710°C but in practice its application is limited to temperatures about 1 250°C, when devitrification commences. The difference between fused silica and ordinary silica (dinas) refractories is that the former does not have a clearly distinguishable crystal lattice. Powdered silica is sometimes used for certain refractory applications; it is obtained by calcining silica gel at temperatures below 1 000°C. These gels are obtained by decomposing silicon fluoride, $SiF_4$, with water, or alkaline silicates with solutions of hydrochloric acid.

Fused silica is very acid resistant and has useful electrical insulating properties. The material is used to prepare spalling-resistant containers and tubes and also mercury-quartz lamps (fused quartz transmits visible light and ultraviolet rays). Since it has a very low coefficient of thermal expansion it is highly thermal-shock resistant.

Slip-cast quartz ceramics, articles made from fused glass, suitably ground and fabricated by aqueous slip casting techniques, have been proposed as substitutes for fused glass articles made by the usual fusion-casting methods. It is claimed that quartz ceramics possess the same basic properties as fused glass, except optical properties. They are very spalling resistant, have low thermal expansion factors and are now finding use both as refractories and in engineering design applications. Specimens can be made by slip casting and firing at temperatures between 1 200 and 1 300°C with soaking periods of around 1 h to give apparent densities of $1 \cdot 9$–$2 \cdot 20$ g/cm$^3$. It is found that the strength of these quartz ceramics, determined under short-term loading, gradually increases with a rise in temperature up to 1 100°C. The gas permeability of the materials having an apparent porosity of less than 8 % is very slight. However, the materials are not always vacuum tight. Quartz ceramics begin to volatilise in vacuum at temperatures of 1 200–1 300°C. According to Popil'skii *et al.* (1971) the electric conductivity is approximately the same as that of fused silica and hardly alters with density in the porosity range 1–10 %.

## REFERENCES AND BIBLIOGRAPHY

Chang, L. Y. and Phillips, B. (1969). Phase relations in refractory metal-oxide systems, *J. Amer. Ceram. Soc.* **52,** No. 10, 527–33.

Gitzen, W. H. (1970). 'Alumina as a Ceramic Material'. American Ceramic Society, Ohio, USA.

Popil'skii, R. Ya. *et al.* (1971). *Ogneupory*, No. 4, 45.

Rice, R. W. (1969). CaO, fabrication and characterization; properties, *J. Amer. Ceram. Soc.* **52,** No. 8, 420–36.

*Chapter 17*

# Non-oxide Refractories

The most important refractory material in this class is silicon carbide since it is used both as a 'bulk' refractory for furnace construction, kiln furniture and crucible making and as an engineering material (technical ceramics). The other compounds included in this category are the other carbides, the silicides, nitrides, borides, phosphides and sulphides. Many of these materials are produced and used in very limited quantities because of their high cost and difficulty of fabrication. Their technology is largely treated within the discipline of the powder–metals specialist.

These non-oxide refractories have certain specific and valuable properties. They invariably have high melting points and are very hard, possessing useful electrical, chemical and magnetic properties. Because of their great hardness and strength they are often used for making hard alloys (for tool making). Many of them are used as the refractory constituents of cermet compositions (*see* below). They are finding use in gas turbines, guided missiles and nuclear engineering where properties such as thermal expansion, creep, etc. are important (*see* Figs. 17.2 and 17.3).

## CLASSIFICATION

Various systems of classifying these materials have been evolved and continuous research is being carried on into the crystal structure, the types of bond, and other physical and chemical characteristics. One useful classification method is to divide the non-oxide refractory compounds into: (1) metalloid, (2) non-metallic, and (3) mixed or intermediate.

The first includes, for example, compounds of zirconium, molybdenum, cobalt and nickel with carbon, boron, nitrogen, silicon, phosphorus and sulphur: the carbides, borides, nitrides, silicides, phosphides and sulphides of metals.

The distinguishing features of group (1) compounds are their metal-like electrical conductivity, high thermal conductivity and thermal expansion factors. They differ from metals in being mechanically weaker,

Fig. 17.1   Special ceramics and refractories being fired at a Royal Worcester factory
in Glamorganshire. *Photo:* Courtesy Carborundum.

much harder and by having lower thermal-shock resistances and much
higher melting points.

Group (2), the non-metallic non-oxide refractories, are formed from a
combination of non-metals, for example boron, nitrogen, carbon and
silicon, giving boron nitrides, silicon borides, silicon carbides, etc. The
most important properties of these compounds are high electrical resistance, low thermal conductivity, high hardness and spalling resistance,
low thermal expansion and good slag resistance.

Group (3), the mixed compounds, have properties that are intermediate between those of groups (1) and (2). The main materials here are aluminium and beryllium nitrides, carbides and borides.

FIG. 17.2. Baudran apparatus for determining the thermal expansion of refractories. *Photo:* Courtesy British Ceramic Research Association.

**Cermets**

Within this class of non-oxide refractories we may also conveniently include combinations of metals and ceramics, known as cermets, although some of the ceramic constituents are frequently oxides. The constituents of these complex materials are usually specified in the name or description of the cermet *e.g.*, TiC–WC, etc.

Fɪɢ. 17.3.   Equipment in use at the British Ceramic Research Association for measuring the creep under compression of materials at high temperatures. *Photo:* Courtesy British Ceramic Research Association.

TABLE 17.1

*Melting Points of Non-oxide Refractories, °C*

| Metal | Compound | | | |
|---|---|---|---|---|
| | Carbides | Nitrides | Borides | Silicides |
| Titanium | 3 140 | 2 950 | 2 980 | 1 540 |
| Zirconium | 3 530 | 2 980 | 3 040 | 1 700 |
| Tantalum | 3 880 | 3 090 | 3 100 | 2 200 |
| Vanadium | 2 830 | 2 050 | 2 100 | — |
| Tungsten | 2 600 | — | 2 300 | 2 165 |
| Silicon | 2 700 (decomposes) | — | — | — |

## SILICON CARBIDE REFRACTORIES

A wide range of silicon carbide, SiC, refractories is produced in various qualities. The compound SiC is not found naturally and has to be synthesised as a crystalline material in electric resistance furnaces using sand and coke. The carbide-forming reaction is quite complex and in order to eliminate harmful impurities such as ferric oxide and alumina, purifying additives such as salt are incorporated and this complicates the reactions even further. The basic reaction can be represented by:

$$SiO_2 + 3C \rightarrow SiC + 2CO$$

sand     coke     silicon     carbon

carbide     monoxide

Commercially the two main classes of silicon carbide refractories are (1) clay-bonded, and (2) self-bonded.

TABLE 17.2

*Properties of Silicon Carbide*

| | |
|---|---|
| Density | $3 \cdot 12$–$3 \cdot 22$ g/cm$^3$ |
| Mohs hardness | $9 \cdot 0$–$9 \cdot 5$ |
| Thermal expansion factor at 1 000–2 400°C | $5 \cdot 68 \times 10^{-6}$ |
| Thermal conductivity, kcal/m h deg: | |
|     at 500°C | 56 |
|     at 875°C | 36 |
|     at 200–1 400°C (recrystallised silicon carbide) | $0 \cdot 04$ |
| Microhardness, kg/cm$^2$ | 3 000–4 700 |

In clay-bonded materials the weak constituent is the clay (aluminosilicates) and during service, especially in highly oxidising atmospheres, the clay tends to exude from the grains of silicon carbide, thus weakening the structure. The manner in which clay is affected by heat and slag during service is complicated and recent research has led in many instances to the elimination of the use of clay and the development of so-called self-bonded silicon carbide refractories.

The biggest disadvantage of silicon carbide 'bulk' refractories is that when they are used in oxidising conditions the silicon carbide 'burns away'. In order to enhance the oxidation resistance, in the past manufacturers increased the clay content, hoping to produce a denser and more heat-conducting product. However, the increased vitreous phase resulting from the reactions in the clay during firing reduced the thermal conductivity and increased the tendency to failure under hot loading. In the manufacture of clay-bonded silicon carbide refractories, therefore, it is essential to use the optimum quantity of clay which is found to lie within the range

3–10%. The grain-size distribution of the silicon carbide is also very important and must be carefully controlled for best quality products.

After the silicon carbide grain and clay (sometimes in the form of a slip) have been blended, the semi-dry bodies are fabricated on toggle or friction presses, by pneumatic tamping, manual tamping and other methods, depending on the size and shape of the articles. The green articles are then fired in tunnel or periodic kilns.

### Silicon nitride-bonded silicon carbide

Sometimes the clay bond is replaced by a silicon nitride bond, which like the main material is produced artificially. It has the formula $Si_3N_4$ and is used in the form of specially milled additions in order to make a highly effective bond for the silicon carbide grain. Most of the disadvantages of clay bonds are eliminated by using the silicon nitride type of bond which exists in two forms, the $\alpha$- and $\beta$-varieties. Commercial refractories usually contain both types of compound. The refractoriness-under-load of nitride-bonded silicon carbide refractories is considerably higher than that of clay-bonded refractories; for example, using the same type of grain this property can be increased from 1 530 to 1 800°C. The use of the nitride as the bond also increases the density and reduces the gas permeability compared with even pure silicon carbide products. This means that the oxidation tendencies are also reduced. Nitride-bonded refractories are not readily wetted by metals and have excellent acid resistance. They do not resist alkalis.

### Self-bonded silicon carbide

When silicon carbide is subjected to oxidising conditions one of the products is silicic acid which forms a bond together with other mineral impurities in the material. By forming a film on the surfaces of the silicon carbide grain at elevated temperatures this product can cause self-bonding of the material without the need for adding any other agents, such as clay or nitrides as discussed above. The oxidation of the silicon carbide is essential for the formation of the bond during firing. At elevated temperatures reactions occur leading to the oxidation of the carbide in different ways since each of the constituents present in the material may remain free or combine with oxygen to form various oxides (silicon monoxide or silicon dioxide) and similarly with the carbon: carbon monoxide or carbon dioxide).

### Other silicon carbide products

It is possible to combine silicon carbide with other materials such as graphite and carbon; also chamotte combinations can be employed to obtain a range of materials for various purposes. For example, combinations of silicon carbide, graphite and 9–10% additions of phosphoric acid will yield very strong materials at room and high temperatures. These materials can be used for making crucibles handling up to half a ton of molten alloys based on copper for use in high-frequency electric furnaces. The compositions can be isostatically pressed and shaped by other methods.

### Uses of silicon carbide refractories

Silicon carbide refractories in the various grades discussed above are employed in the metallurgical and ceramic industries and for special engineering purposes. In the iron and steel industry they are used for high-temperature recuperators for continuous steel-casting techniques and in powder metallurgy in the form of self-bonded silicon carbide components for making stamps and dies. Their high abrasion resistance is utilised in the manufacture of rails and sliding blocks. They can also be used in place of metals for cooled water pipes and for non-cooled metal rails. In the ceramics industry silicon carbide kiln furniture is widely used in tunnel kilns and in periodic kilns because of its high thermal-shock resistance, high thermal conductivity and mechanical strength. The chemical engineer makes use of silicon carbide refractories for spray nozzles, for components where metals have inadequate corrosion resistance and for certain abrasive-resistant pump units for handling corrosive liquids, etc.

## NITRIDES

Nitrides, of which boron nitride is the most important to the refractories technologist, are not used as 'bulk' refractory materials. Nitride-bonded silicon carbide has already been mentioned; nitride bonding is an important aspect of the physical chemistry of refractories science. Boron nitride is a white soft powder with a hardness of about 2 on the Mohs scale. It melts at 2 350°C. It is very resistant to chemicals in reducing and neutral atmospheres, has a high thermal conductivity and consequently a high thermal-shock (spalling) resistance. Because of its outstanding dielectric properties it maintains its insulating capacity up to very high temperatures, approximately 1 800°C in non-oxidising conditions. In addition to the white boron nitride (also known as 'white graphite' because of its plate-like structure, similar to graphite) there is also a cubic structured boron nitride known as *borazon* which is very hard and sublimes at a very high temperature. The fabrication of boron nitride refractory components is complicated because the material does not sinter very well and brittle structures result. Hot pressing using graphite moulds at temperatures of about 1 700–1 800°C and pressures up to 450 kg/cm$^2$ is used for fabrication purposes. The density achieved is about 90% of the theoretical. Borazon has a hardness approximating to that of diamond but its specific heat is double that of diamond and it is an excellent electric insulator. One of the important advantages of borazon is its oxidation resistance up to 2 000°C, while diamond resists oxidation only up to about 800–850°C.

*Nitride bonding* has been mentioned above and in the section on silicon carbide refractories. The technique can also be used for producing insulating silicon carbide materials by the foam method. In this case the nitride bond results from the reaction between a mixture of silicon nitride, in

Fig. 17.4. Silicon carbide refractories are used for their high thermal conductivity and strength properties inside tunnel kilns. *Photo:* Courtesy Carborundum.

the α- and β-modifications, and nitrogen. The nitride-forming reaction commences at about 1 200°C and is complete at 1 450°C. The rate of nitride formation in foamed refractories is much greater than in dense specimens. The product has a homogeneous phase composition, which gives it an advantage over foamed silicon carbide with a traditional bond (produced in carbon fillings and packing). Nitride-bonded foamed insulating products have a high thermal-shock resistance and structural strength at elevated temperatures. Foamed silicon-carbide insulating products can be used for high-temperature applications in non-oxidising conditions when a high spalling resistance is essential.

*Silicon oxynitride* is also of interest in the bonding of certain special-duty refractories, particularly with reference to the bonding of silicon carbide as a means of improving the high temperature properties. It is found that the corrosion resistance to molten metal slags and other agents can be greatly increased by replacing conventional bonds in silicon carbides by silicon oxynitride.

Ordinary silicon nitride, $Si_3N_4$, develops from the action of amorphous silicon with nitrogen above 1 000°C. At a temperature of about 1 900°C silicon nitride decomposes. Like the boron nitride it is chemically resistant to most molten metals and fused salts. Various other nitrides have been produced commercially or experimentally, including aluminium nitride, molybdenum nitride, etc.

## BORIDES

Borides are compounds obtained from boron and the metals mainly of the transition group. Most borides have very high melting or decomposition points. They are also distinguished by great hardness and excellent spalling resistance. Borides do not tend to volatilise very readily which means they can be used at temperatures above 2 300°C. They are, however, sensitive to oxidation, and their valuable refractory properties can be utilised only in reducing conditions. Borides have a high electrical conductivity and a positive resistance factor (high electrical resistance) as a result of which they are used for making heating elements. Zirconium boride, hafnium boride, titanium boride and others, are better electrical conductors than the metals used in their composition. Products made from borides are usually fabricated by pressing either by the isostatic or conventional methods, followed by sintering. Hot pressing has also been used. Because of the high cost of materials and fabrication, borides are not used as 'bulk' refractory materials and their applications are usually restricted to special engineering and chemical engineering applications. They are very resistant to the action of most acids, including HF. In molten alkalis, carbonates and sulphates they decompose. Borides frequently form the basis of the class of materials known as cermets; for example, zirconium boride with a complex metallic bond system containing chromium and molybdenum has very high refractory properties. This cermet group will resist the action of molten metals such as aluminium, tin, brass, copper and lead at temperatures ranging from 400 to 1 150°C. Cermets based on titanium boride and 5% molybdenum show almost no reaction with molten lead, cadmium, bismuth and tin.

## SILICIDES

Silicides are compounds between silicon and metals, commonly from groups 4–6 in the periodic table. Many silicides are known but most of them are limited to very specialised uses. The commonest material is molybdenum disilicide which is used in special-refractory engineering. Other silicides of practical importance are zirconium silicide, $ZrSi_2$, tantalum silicide, $TaSi_2$, and tungsten silicide, $WSi_2$.

Molybdenum disilicide has a low electrical resistance and is widely used in making heating elements. These can operate at temperatures around 1 680°C for several thousand hours. Above this temperature a protective glassy film tends to develop on the surface of the elements and the molybdenum disilicide is oxidised. Molybdenum silicide coatings are also important in high-temperature engineering. Attempts have been made to use the material in gas turbine engineering, guided missile design and in nuclear engineering.

## OXIDATION RESISTANCE OF NON-OXIDE REFRACTORIES

As might be expected one of the important differences between oxide and non-oxide refractories is that the latter have an affinity for oxygen during heating, and this, despite their valuable properties, can limit their use, especially at very high temperatures. Various techniques are employed to overcome this weakness, including surface treatment, impregnation, etc. However, the oxidising tendencies of these materials must always be taken into account in any application. The oxidising mechanism can be used as a means of classifying non-oxide refractories. For example, when nitrides and carbides are oxidised gaseous oxides of carbon and nitrogen, which break up the oxide film on the surface of the articles, are formed, together with the corresponding oxides of the metals. This means that these refractories have a relatively low slag and metal-scale resistance. When borides are subjected to oxidation, in addition to the boric oxide volatilised, the reaction products include borates which tend to form dense protective layers, making borides more resistant than nitrides and carbides to slagging. The oxidation of silicides, such as those of molybdenum and vanadium, results in the development of a thin silica film or solutions of silica and silicides which prevent further oxidation. At temperatures above 1 700°C the silica film 'crawls' in the manner of a badly fired ceramic glaze on the refractory due to the action of surface tension and exposes fresh layers of silicides which may be rapidly eroded or corroded by the environment.

Surface reactions and the development of protective films, either by applying coatings before use, or in the service conditions for a particular material, constitute an important branch of refractory materials science. The manner in which fused metals and compounds, including slag compositions, react with refractories, cermets, and refractory metals is very complicated and represents an important field of current research.

## PRODUCTION TECHNIQUES

In addition to conventional powder-metallurgical pressing of blanks, followed by sintering, simultaneous pressing and sintering (hot pressing), the refractories technologist employs reaction sintering for the fabrication of non-oxide refractory materials. Reaction sintering consists in combining the formation of the refractory compound and the sintering process. The process is used, for example, for making articles of silicon nitride, boron nitride containing graphite, aluminium nitride, and silicon carbide.

## PROPERTIES AND USES OF NON-OXIDE REFRACTORIES

As mentioned above, non-oxide refractories find their main uses in chemical engineering, rocket and missile design, for special engineering purposes, and in laboratory technology. Examples of the uses of these materials include thermocouples, thermocouple sheaths, tubes and piping, electrodes, heating elements, and crucibles for the melting and processing of special alloys and metals.

Some non-oxide refractories are particularly suited for fabricating thermocouple sheaths for measuring high temperatures and for measuring the temperature of molten metals, glasses, etc. Zirconium boride, $ZrB_2$, may be used for such a purpose and sheaths have been used for measuring the temperatures of molten iron in blast furnaces. The advantage of this material for this purpose is its low gas permeability. Gas admission often causes failure of the thermocouple due to corrosion and crystallisation of the noble-metal thermocouple wires. The design of the thermocouple

TABLE 17.3

| Material in sheath | Used for | Temperatures |
|---|---|---|
| $ZrB_2$ | Blast furnaces (11 h dwell) Open-hearth (2–3 h dwell) | 1 500–1 600°C |
| Composite: $ZrB_2$ and $Al_2O_3$ | Open-hearth (2 h dwell) | 1 550–1 630°C |
| Nitride-bonded silicon carbide and refractory steel | Molten aluminium electrolyser (>100 h) | 950–980°C |
| Silicon nitride, silicon nitride and boron nitrides, nitrides and carbides of silicon | Molten brasses, bronzes, etc. | Various |

sheath is very important and several versions for which the elimination of the above drawbacks is claimed, have been evolved. Sometimes multiple sheaths are made up with different refractory materials, for example, zirconium diboride, alumina and alundum.

In the measurement of temperatures in aluminium electrolysers, aluminium nitride has found an important application in the manufacture of couple sheaths because it has hardly any reaction with molten aluminium or cryolite, one of the raw materials used for smelting aluminium.

## BIBLIOGRAPHY

Frantsevich, I. N. (1970). 'Silicon Carbide'. Consultants Bureau, New York.

Gladyshevskii, E. I. (1971). 'Crystal Chemistry of Silicides and Germanides'. Stroiizdat, Moscow.

Henisch, H. K. and Roy, R. (1969). 'Silicon Carbide—1968'. Pergamon Press, New York.

Samsonov, G. V. (1969). 'Nonmetallic Nitrides'. Metallurgiya, Moscow.

# Chapter 18

# Concretes and Castable Refractories

In many furnace linings the first part to give way to the penetrating action of molten metals and slags is the brickwork joint. The advantage of using refractory concretes or ramming bodies that can be forced into furnace linings to make monolithic structures is that no joints are formed, and this eliminates a possible source of weakness in service. The need for scarce skilled bricklaying labour is also eliminated, or greatly reduced. The use of castables in furnace construction also facilitates the use of mechanised methods of building and repair.

By selecting the quality of aggregate or filler and the type of cement (usually high-alumina varieties) it is possible to compile castable compositions to suit the particular application. As with fabricated and fired refractories, made by the normal technology, a range of structures can be obtained, from dense to highly porous. The degree of porosity is partly determined by the structure of the aggregate (chamotte, grog), and partly by the grain-size distribution and method of preparation for the whole placement.

As in ordinary building concretes, refractory concretes and castables consist of a cement or bonding agent and an aggregate. Various types of additive are also used to confer certain properties, such as workability (plasticity aids) and sometimes to accelerate or retard the setting process. Refractory concretes can be applied by shuttering, dry gunning and wet gunning.

The firing of refractory castables takes place after they have been built into the furnace. In this respect they differ from conventional fired bricks, slabs and shapes. The method of use depends on the quality and application. For example, common aluminosilicate concretes can be mixed in the site, using fireclay chamottes, aluminous cements and the necessary quantities of water, in much the same way as non-refractory concretes are mixed and placed. In other cases the dry mix may be delivered in sealed drums or bags and this needs the addition of water; in other instances the castable may be blended with water at the production factory and delivered in air-tight drums ready for use.

165

## PROPERTIES OF CASTABLES

The properties will depend mainly on the chemical and mineral composition of the constituents, as with fired refractories, and on the manner in which the placed material is treated. Special care must be taken with high-clay compositions to ensure that the firing is oxidising in order to remove organic matter. Linings will otherwise have black cores and upon the first serious onslaught from slag or molten metal will fail in service.

It is common practice to ram the metal housings of frit and enamel-melting kilns used in the pottery industry with plastic refractory materials composed of chamotte and fireclay or kaolin. If the heating of the kiln prior to the first melting is not done properly the lining soon fails because of the penetration of melt into the weak, unfired 'core' between the metal housing and the hot face of the lining.

The development of the structure (porosity and strength) will also depend on the heating-up procedure.

## CLASSIFICATION OF CONCRETES, MORTARS AND CASTABLES

A very wide range of compositions and forms is in use. The classification below is tentative and merely indicates the purposes and areas of use.

1. *Mortars and washes*, used after slaking with water for bricklaying and joint making in refractory brickwork.
2. *Ramming bodies.*These are plastic, semi-dry compositions used for building up monolithic structures by hand with pneumatic tools, or by other tamping methods.
3. *Refractory concretes.* These are very similar to ramming bodies except that they contain a hydraulic or air-setting bond (cement) such as Portland, aluminous cement, or aluminophosphate bonds. Sometimes compositions such as ethyl silicate are used. These concretes may also be phosphate bonded (*see* below).
4. *Fettling, repairing compositions.* These are used for laying and repairing the hearths, bottoms and other parts of furnaces, *e.g.* open hearths in the steel industry.
5. *Guncrete and coating bodies.* These are used for repairing and protecting existing structures. They are formulated especially for application by spray guns and can be used internally or externally.
6. *Metal coatings.* Refractory compositions are sometimes designed for coating metals in order to protect them against heat and oxidation.

As with fired refractories, coatings and castables may also be classified according to their chemical compositions as well as with respect to the purpose for which they are formulated. The type of bond is yet another criterion by means of which refractory mortars and concretes are classified, for example: (1) hydraulic; (2) water glass and other mineral adhesives; (3) organic adhesives such as sulphite lye, coal-tar pitch, etc; (4) chemical bonds, for example, phosphates derived from adding phosphoric acid to the mix.

## REFRACTORY MORTARS

These are mixtures of refractory constituents, bonding clays and other agents which after blending with water to form a paste are used to build up the brickwork structure. The chemical and mineral composition of the mortar must be selected to suit the compositions of the fired refractory bricks and blocks. Thus, the composition may be clay-grog, silica, aluminosilicate, chromite, carbon paste, magnesite–chromite, etc., depending on the composition of the brickwork. Since the aim is to provide the mason with a material for producing very thin joints, it is obvious that the preparation of these mortars for high-quality linings is of the utmost importance. The particle-size distribution and the ratios of aggregates:bond:water must be clearly selected. The essential properties to be considered in the preparation and use of refractory mortars (and also concretes) include the water-retaining capacity, the chemical composition, the grain-size distribution, the drying and firing shrinkage, the refractoriness, the bonding capacity, and the gas permeability.

Sometimes it is necessary to lay the brickwork 'dry'. This involves careful placement of the refractory pieces in the lining without using mortar and then filling in the joints or other gaps with fine powder of the same composition as the products. Sometimes metal plates are used to separate the bricks and shapes and during firing these plates fuse and react with the bricks to form a more or less monolithic structure. The laying of *carbon blocks* involves the use of a paste made from ground coke and tar which develops a coke-like structure when fired to high temperatures.

Mortars prepared for the laying of *insulating* shapes, made by the foaming method or by the combustible-additives method, must be of the highly adhesive variety; for example, when fireclay or high-alumina insulation is being constructed it is common practice to add about 2% water glass to the mortar. This increases the bonding capacity and reduces the moisture-extraction rate from the brick and helps to build up a sound structure. Hot-face insulating bricks are usually laid with mortars made from kaolin chamotte (calcined china clay).

Blocks of *diatomite* used for insulating purposes can be bonded in brickwork construction with mortars containing about 80% diatomite,

10% starch and 10% plastic clay, or similar compositions. Silica insulating brick can be bonded with mortars containing 90–93% silica scrap and 10–7% high-quality fireclay.

*Air-setting mortars* are those which contain some kind of adhesive or bonding agent that acquires its set when exposed to the atmosphere. For example, additions of water glass to clay–chamotte compositions constituted as mortars, will cause the resulting mortar to set hard in air. The use of water glass and similar fluxing agents is prohibited if the linings are to operate at very high temperatures. It is then customary to use simple grog–clay or chemically bonded refractory linings. The fireclay air-setting mortars have a much lower refractoriness than similar mortars not containing the water glass.

## RAMMING BODIES

The compositions of ramming bodies vary as widely as those of fired refractories, for example high-alumina, silica, fireclay, magnesite–chromite. The purpose of using ramming bodies is to eliminate mortar joints in the structure; *i.e.*, they are used to form a monolithic lining. As with concretes and mortars, ramming bodies consist of mixtures of refractory aggregate and bonding materials used in the moist or dry form for lining fabrication. The shaping of the complicated parts of the lining is greatly facilitated by the use of ramming bodies in instances where brickwork construction is difficult. One possible disadvantage with the use of ramming bodies, particularly when they are very wet, is the longer time needed for drying out the linings after the ramming has been completed. Since the conventional grog–clay mixtures may exhibit substantial shrinkage during drying and firing it is sometimes useful to incorporate *expansive additives*, e.g. kyanite, to compensate for the high shrinkage. This helps to produce a volume-stable lining.

With the scarcity of skilled labour for brickwork, ramming techniques have been widely developed in the refractories industry. For example, high-alumina shrink-free ramming bodies consisting of coarse china-clay grog with grain sizes of 5–0·5 or 3–0·5 mm and a proportion of fines (grain sizes less than 0·08 mm) together with fused alumina or sillimanite and other grogs, are extensively used for making high-temperature linings. Such compositions are suitable for service at temperatures up to 1 650°C. Other materials suitable for making ramming bodies include siliceous compositions, chromite, magnesite–chromite, etc. Where the composition does not contain any 'fusible' constituent which will develop moderate temperature bonding, it is necessary to add a material such as orthophosphoric acid which will develop strength at temperatures around 500–750°C. These ramming bodies are prepared in runner mills, for example, and are used in the moist state for fabricating the linings of induction furnace crucibles for smelting aluminium alloys.

Chromite ramming bodies are made from chromite ore of certain grain-size compositions and water glass or clay bond.

Ramming bodies suitable for making carbon linings are made from coke, anthracite, and carbon scrap aggregate with a tar bond prepared from pitch and anthracene oil. As with ordinary carbon blocks the main disadvantage of these materials is their low resistance to oxidation (they burn up in oxidising atmospheres), but they have very high refractory properties in neutral or reducing atmospheres, including constant volumes, good spalling resistance, excellent thermal conductivity and slag resistance. These ramming compositions are often used to fill in the gaps between carbon blocks in blast furnaces and for making the linings of electric arc furnaces.

## GUNCRETE COATINGS

As the name indicates, these are concrete compositions that are applied pneumatically by means of spray guns. A very wide range of compositions has been developed for different purposes. As with the preparation of any material for spraying, the particle-size distribution and rheological properties are very important to ensure efficient application. Guncretes are

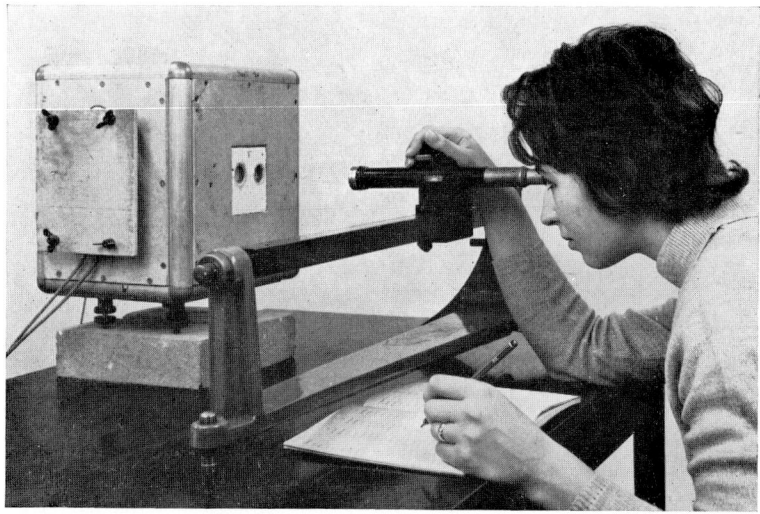

FIG. 18.1. Furnace and cathetometer being used for measuring the movement of specimens during heating. Such movement is important both in fired and unfired refractory linings. *Photo:* Courtesy British Ceramic Research Association.

applied to the faces of refractory linings to prevent early wear due to the action of slags, molten metals, gases and other corrosive agents. They are also used for repairing damage while the furnaces are still operating or when the furnaces have been stopped (dry gunning). Relatively thin

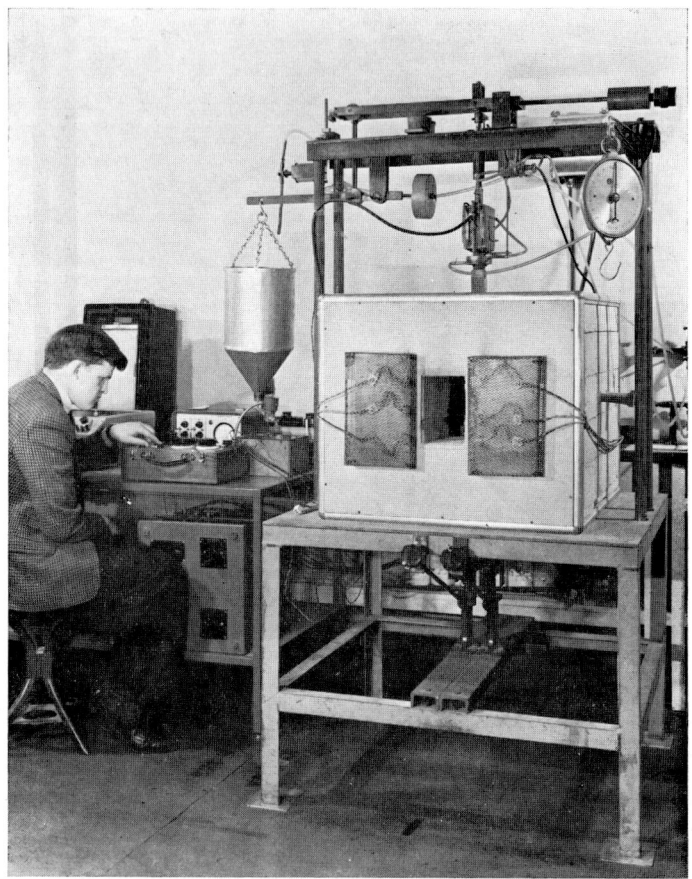

FIG. 18.2. High-temperature modulus of rupture testing equipment. *Photo:* Courtesy British Ceramic Research Association.

coatings (not more than 5–6 mm) only can be applied by the guncreting technology. The problems involved are ensuring that the material fed to the guns has a consistent composition and grain-size distribution. In some equipment the dry refractory powders and the moistening agent (water or solution of bond) are mixed in the mixing head of the gun.

Thus, the materials applied in this way may have an irregular moisture content. With other equipment the bodies are prepared, moistened and then fed to the gun; they then have a more regular moisture content and the results are much more satisfactory. The essential properties of any compositions prepared for guncreting are excellent bondability to the refractory base to which they are applied, minimum drying and firing shrinkage and a high resistance to mechanical and chemical action (Figs. 18.1 and 18.2.)

## VIBRATION CASTING

Recently some of the disadvantages of the existing methods of applying refractory concretes to furnace linings (dry and wet gunning, and shuttering) have been removed by the development of a patented vibration-casting technique (Vycast System). This involves constructing furnace linings as a series of panels in light steel moulds. The mould side-plates are bracketed to the steel casing of the furnace on screwed studs which are welded to the casing. The concrete, prepared with the minimum practical water content, is put in the mould and is then consolidated with a rotary air-operated vibrator (1 200 vibrations per minute) which is fixed centrally to the external surface of the mould. A pulley system than raises the front panel of the mould, exposing the slab of concrete, which is by now solidifying. Mould filling proceeds, the front panel is vibrated and raised, and eventually the slab of concrete reaches the desired height.

Furnace linings from 7·5 to 80 cm thick have been formed with this vibration-casting method. Several equipment assemblies can be used simultaneously on one lining to speed up the job. The equipment can be adapted for difficult sections of the lining and hand-held moulds are available to match tapering sections, circular parts and triangular closing pieces. In the lining of circular structures, *e.g.* conditioning towers and some zones of blast furnaces, where concrete is used in areas that are prone to severe abrasion, it is possible to employ segmental moulds, while complete circular moulds can be used for some vertical cylindrical ducts.

One mould needs only one operator and a team of workers each handling one mould can considerably accelerate lining work compared with the more cumbersome shuttering technique. Another advantage over shuttering is that the placed concrete panel is opened up for inspection soon after is has been formed, whereas with shuttering, voids, cavities and cracks are revealed only after the concrete has solidified. Patching up must then follow.

The shuttering of furnace parts which are difficult to get to often leads to crevice formation near metal and clay anchors (metal anchors are jagged pieces of steel welded to the furnace housing, by means of which the concrete holds on to the lining). This is easily understood because the operator cannot get to these difficult corners to tamp the concrete in place. Vibrating the concrete, which is often 'sluggish', as in the Vycast method, eliminates this fault.

Speed of placement is an important factor in furnace construction and repair. Vibration placement using the above method is much faster than guncreting. According to the inventors of the method, where a gunning team would place 15 tons of concrete a day, an equivalent Vycast team would place 25 tons a day.

### Uses of vibration-casting system
It is apparent from the above description that vibration casting is applicable to most large furnaces whose linings can be economically constructed by mechanised methods. Actual applications include the boiler side-walls of a high-pressure boiler; part of a soaking pit wall in a steel mill; linings of burner tubes in a carbon-black factory when the tube was too small in diameter for gunning, and too long for shuttering; and exhaust duct lining in a carbon-baking furnace.

A satisfactory structure in a soaking pit, for instance, could be built up from a layer of high-duty refractory concrete (say high-alumina, or sillimanite type) backed by a layer of lower duty material such as common chamotte (fireclay).

Suitable metal anchors used to retain the furnace linings are 'Y' types of 30 cm square pitching, but other types can also be used. In all kinds of refractory concrete placement anchors must be used to ensure that the huge monolithic lining does not collapse into the furnace, as might well happen with a sudden change in the heating schedule, leading to thermal shock on a large scale. For the soaking pit lining mentioned above metal anchors can be inserted through the backing, and serrated fireclay blocks are fixed to the anchors to retain the high-duty face lining. A single mould is used to cast the two concrete layers.

*Particle-size distribution*, as discussed above, is an important criterion in the fabrication of any refractory lining, not only in the placing of the concrete but also as regards subsequent performance, particularly wear resistance. One disadvantage of dry guncreting is that the filler (grog, chamotte) particle sizes must be restricted to the lower range, since the powders have to travel along hosepipes. Such fine particle sizes prevent the development of high-density structures with good abrasion resistance. Vibration casting allows large aggregate particles to be used (up to 3·5 cm) with a consequent improvement in the abrasion resistance of the set concretes.

## TYPES OF BONDS

The bonds used in preparing refractory concretes, mortars, ramming bodies and coatings include Portland cement; aluminous cement; barium cement (for radiation-resistant concretes); water glass (potassium or sodium silicates); periclase cements (magnesia cement); phosphates, and other chemical bonds such as ethyl silicate, etc.

High-alumina (aluminous) cements are rapid-setting hydraulic bonding materials, like Portland cement, produced by grinding clinker obtained by the normal cement-firing process, but employing high-alumina materials. They usually contain 34–52% alumina, ensuring a high concentration of calcium aluminates. Good quality aluminous cement begins to set after about 25 min and is completely hard after 10–12 h following the addition of water. High-alumina cement (calcium dialuminate) consists of material with at least 75% $Al_2O_3$, and a PCE of about 1 800°C. It begins to set in 45 min, and its setting is complete in about $3\frac{1}{2}$ h.

Sodium silicates (water glass) are liquid or powdered materials with the general formula $M_2O.nSiO_2$ where M can be sodium or potassium, and n is a factor which determines the silica modulus of the material. Sodium silicates have a silica-soda modulus in the range 2·4–3·0. It is important to specify the silica modulus in using water glass for refractory-concrete and mortars. Since it is a glass, sodium silicate behaves as a fluxing material when used in conjunction with refractory materials and its use is therefore limited to relatively moderate temperatures.

Periclase cement is obtained from magnesite, finely ground, and mixed with water or solutions of magnesium sulphate and magnesium chloride. When set it forms hydrated magnesium oxide which is a bonding agent. It is used for bonding magnesite refractories.

Phosphate bonds are finding increasing use for unfired refractories and refractory concretes, castables and similar compositions. Phosphate-bonded refractories have high mechanical strengths at temperatures between 400 and 1 050°C, that is, precisely when the strength of other bonds such as magnesia and water glass is very low. The advantages of phosphate-bonded refractories also include the ability to resist the wetting action of molten metals.

The simplest way of using the phosphate bonding technique perhaps is to add orthophosphoric acid to a fireclay body (chamotte) in order to strengthen and enhance the sintering of the body and to improve the crystallisation. This means that the resulting material can be used as an unfired material for ramming applications. The optimum concentration appears to be 2–3% of 85%-concentrated orthophosphoric acid. The strength rises with an increase in the concentration of alumina in the material, using fine fractions of clay.

The phosphate bonding technique can also be used to make unfired forsterite refractories. Products containing magnesium–phosphate bonds have a high strength, even some considerable time after pressing. After 24 hours storage the strength of these articles (made from dunite and magnesite) may rise by about six times, and the products based on the synthetic briquette by 4–12 times. The strength of the refractories is somewhat increased during further storage over a period of seven days. Heat processing for two hours at 350°C hardly affects the strength and may even somewhat increase it. Magnesium phosphate bonds increase the strength of forsterite products during heating, especially in the range 500–700°C. The adhesive capacity of magnesium phosphate bonds allows them to be used in the production of unfired forsterite products and concretes.

One possible serious defect with the use of phosphate bonds in chamotte bodies is bloating during heating. This has been traced to the evolution of hydrogen when the acid reacts with metallic impurities. The extent of the bloating depends on the quantity of the impurities. Raising the temperature of the chamotte body containing the phosphoric acid to 80°C tends to increase the volume by bloating 3–5 times. This weakens the structure and makes these materials possibly dangerous in use. Employing partly neutralised alumino-phosphate bond reduces the total bloating volume by 20–30% and reduces the maximum bloating rate by a factor of 2–3. [S. R. Zamyatin, *Ogneupory*, No. 11, 44–9 (1968).]

Silicon carbide mixtures can also be bonded with alumino-phosphates. These materials after heating have a very high strength and refractoriness under load. Alumino-phosphates in amounts of 18–20% with a pH of 1·8–2·0 have been recommended for bonding silicon carbide materials. The lining of boiler work is often done with silicon carbide ramming bodies bonded with water glass and clay. As the temperature rises the material loses its strength due to the fusion and loss of the water glass. Thus, it may be possible to use alumino-phosphate bonding to eliminate this kind of fault. The process of bonding various materials with alumino-phosphate is very complicated but can be said to be due to the formation of acid phosphates and their transformation into basic phosphates. The process is critically influenced by the water content, the temperature and the drying rate, the acid concentration and other factors.

Magnesite can also be phosphate bonded. Cements with relatively rapid setting characteristics slaked with solutions of orthophosphoric acid salts have a moderate strength in the range 800–1 200°C. However, cements with long setting periods slaked with solutions of polyphosphates have a much higher mechanical strength at the same temperatures. Thus, it is possible to use phosphate bonding for increasing the refractoriness and strength characteristics of magnesite materials.

Alumino-phosphate bonds consist of colloidal solutions of alumino-phosphates resulting from the reaction of hydrated alumina and diluted orthophosphoric acid. Magnesium phosphate bonds are prepared in a similar manner, ensuring the presence of magnesia to develop the necessary structure. They are used for making basic refractory concretes and ramming bodies.

## FIBRE-REINFORCED CONCRETES

The use of fibreglass in constructing a wide variety of articles, from boats to vaulting poles used in athletics, is now well known. The use of fibres made from other materials such as steel, carbon and asbestos as reinforcing agents in the fabrication of composite materials is not so well known but is becoming an important part of materials technology, especially for critical components, *e.g.* carbon in aircraft and aerospace.

The advantages of composites include enhanced bending strength and the possibility of combining the benefits of several quite different types of structures. The idea of using steel wires and rods in concrete to strengthen it is, of course, the basis of a vast industry. More recently refractories researchers have been investigating the concept of fibre reinforcement for ceramics, polymer concretes and castables, including high-temperature concretes.

Flajsman, Cahn and Phillips (1971), for instance, have reinforced bars of mortar with chopped steel and fibreglass and subsequently impregnated with methyl methacrylate prior to polymerisation. These materials did not have particularly high temperature resistance but the principle may be useful for castable technology in furnace design.

In other recent work, Lankard and Sheets (1971) studied how the addition of stainless steel fibres affected the behaviour of refractory castables based on high-alumina cements. Incorporating 1–2% by volume of wire fibres greatly improves the flexural strength and allows the castable to withstand high stresses even when the matrix has failed completely. This development offers promise for boosting the life of linings and structures where thermal shock and mechanical shock are severe problems.

At lower temperatures (400–800°C) it is possible to use fibreglass reinforcement in all qualities of castable. The use of glass fibre reinforcement has been described for some special application in which there is a need for both thermal and electrical insulation calling for a semi-refractory with good spalling and mechanical shock resistance. A simple technique involves the use of a cardboard tube as the mould and the vibration casting of the alumina slurry. The fibreglass is then worked into the mass using a rod.

## REFERENCES AND BIBLIOGRAPHY

Aukern, A. and Horn, W. (1971). Some properties of polymer-impregnated cements and concretes, *J. Amer. Ceram. Soc.* **54**, No. 6, 282–5.

Dutta, B. M., Pramanik, S. *et al.* (1969). Studies of castable refractories using high purity calcium aluminate cements, *Trans. Indian Ceram. Soc.* **28,** No. 5, 137–43.

Fisher, K. (1969). Chemical bonds for refractory materials, *Proc. Brit. Ceram. Soc.* No. 12, 51–4.

Flajsman, F., Cahn, D. S. and Phillips, J. C. (1971). *J. Amer. Ceram. Soc.* **54,** No. 3, 129–30.

Lankard, D. R. and Sheets, H. D. (1971). *Bull. Amer. Ceram. Soc.* **50,** No. 5, 497–500.

Livovich, A. F. (1966). Properties of pneumatically placed refractory concretes, *Bull. Amer. Ceram. Soc.* **45,** No. 1, 11–15.

Perry, J. D. (1967). Effect of compositional variables on the properties of refractory castables, *J. Canad. Ceram. Soc.* **36,** 45–8.

Petzold, A. and Röhrs, M. (1965). Refractory concretes, in: *Beton für Hohe Temperaturen.* VEB Verlag für Bauwesen, Berlin.

Savioli, F. (1969). Reaction kinetics in monolithic unshaped refractories, *Berichte Deut. Keram. Ges.* **46,** No. 7, 372–4.

Staron, J. (1969). Phosphate bonding in basic refractories, *Berichte Deut. Keram. Ges.* **46,** No. 7, 369–71.

*SECTION III*

# USES

*Chapter 19*

# Use of Refractories in the Iron and Steel Industry

The production of iron and steel consumes about 70% of the total output of the refractories industry. Many steel producers have their own refractories divisions, and all of them carry out routine tests on deliveries of new refractories. Some steel companies possess large research and development departments in which work is being continuously done to produce new types of refractories.

Hot molten metal and slag are the destroyers of furnace linings in the iron and steel industry. They splash, corrode, erode, wash away, dissolve and eventually cause the destruction of what are apparently dense chemically stable and hard materials. The aim of the refractories maker is to produce a lining material which will not react with nor be penetrated by the slag or metal. Ideally he wants to obtain a flaw-free, perfectly inert, thermally shock resistant material. In practice, of course, he must strike a compromise.

## THE BLAST FURNACE—IRON MAKING

In spite of possible future competition from direct reduction processes the blast furnace is likely to remain for many years as the main type of furnace for producing iron. The current trend is towards the erection of larger furnaces; the use of higher temperatures; oxygen enrichment of the blast, and high top pressures. This will mean increasingly critical conditions in refractory service.

The blast furnace is a tapered steel cylinder lined with refractory materials produced in a variety of forms to meet the markedly different conditions in various parts of the structure. The burden, consisting of iron ore, coke and limestone, is fed in at the top of the furnace which may be as high as 36 metres. Preheated air, and, frequently, injected materials such as oxygen, steam and other gases, are forced up the stack through the tuyeres. The coke burns to carbon monoxide and reduces the iron oxide of the ore to metallic iron: the product of the blast furnace, according to the following (simplified) reaction:

$$Fe_2O_3 + 3CO \rightarrow 3CO_2\uparrow + 2Fe$$

The molten iron is tapped at the bottom of the blast furnace and is either delivered directly to the steel-making furnace, or cast and cooled for subsequent processing.

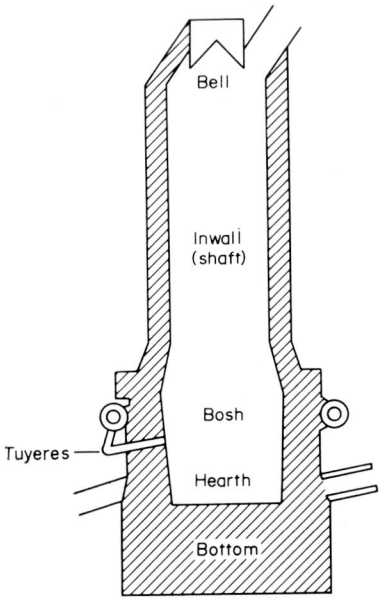

FIG. 19.1.    The main parts of a blast furnace.

The blast furnace operates on the principle of the contraflow exchange of heat and chemical energies. In spite of more than a century's experience with the blast furnace production of iron, very little is known about the temperature distribution in working blast furnaces; this hinders a complete understanding of the precise service conditions of the refractory materials used in the throat, the shaft (known also as the stack and inwall), the bosh, and the hearth: the main parts of the blast furnace (*see* Fig. 19.1).

Modern blast furnaces are now being built to last for as long as 20 years. Careful choice of suitable refractories, together with water cooling of the hottest parts mainly around the bosh and well, give extended campaigns, although frequently unknown deleterious factors are at work and sudden breakdown may occur. An idea of the temperature conditions inside a blast furnace can be obtained from Table 19.1 which shows the gas temperature in the various zones.

The gas leaving the top of a blast furnace contains 10–15 $g/cm^3$ of highly active dust, consisting of particles of ore, coke and fluxes. In some cases the dust concentrations may be as high as 100 $g/cm^3$. The combination of high temperatures, slagging from the dust, and the abrasive action of the burden moving down the furnace, determines the life of the lining. Since these processes operate at varying rates in different parts of the furnace the designer must choose different refractories for different zones.

TABLE 19.1

*Gas Temperatures in the Various Zones of a Blast Furnace, °C*

| Furnace zone | Max. temp. | Min. temp. |
| --- | --- | --- |
| Hearth | 1 900 | 600 |
| Top of bosh | 1 450 | — |
| Bosh | 1 200–1 300 | 960 |
| Shaft: | | |
| bottom | 1 100–1 145 | 805 |
| middle | 925–980 | 540–700 |
| top | 580–620 | 305–400 |

As far as refractory use is concerned a blast furnace can be divided into two main zones:

(a) the upper zone where the most critical part is the shaft;

(b) the lower zone comprising the hearth, the bottom, the bosh and the tuyere region.

Very roughly, the temperature limits in the lower zone are 1 300–1 800°C, and in the upper zone 200–1 300°C. The lower zone has to put up with the action of high temperatures, molten iron and slag and can do so only if the lining is cooled with water. The upper zone must be built to withstand the mechanical action and chemical corrosion from gases and vapours emitted by the burden.

Once a blast furnace is in operation the hearth will be subjected throughout its life to the action of molten iron at temperatures around 1 500°C and pressures of about 5 $kg/cm^2$. Poorly made hearths soon fail: the refractories (carbon or firebrick) float in the molten metal and the furnace operators may be in danger from iron break-out.

The hearth of a blast furnace must be built with special care, for once major repairs are necessary the entire furnace must be dismantled. In addition to the quality of the refractories, the rates of erosion and corrosion are affected by the types of iron being smelted. Foundry pig irons and ferro-silicons need a hotter furnace than ordinary steel-making pig iron. Phosphorus irons which are less viscous also cause faster bottom erosion. The wearing away of bottoms in a blast furnace is a most complex and almost inexplicable process, because it is not easy to take samples from, and examine, the refractories during operation. The bottom works continuously and monolaterally under molten iron and high pressures.

### Furnace lining factors

Effective blast furnace operation is governed by the furnace design and its cooling, the quality and properties of the refractories, the particle size of the charge (iron ore, coke and limestone) and the general working of the furnace considered as a continuous, almost automatically functioning unit. Special attention is paid to the particle size of the burden materials since it is certain that the amount of fine dust in the charge has a critical effect on, for example, the life of the bosh and shaft linings. The use of dense, kaolin-type refractories in the shaft, preventing the penetration of carbon monoxide (*see* below) has greatly reduced the importance of one of the classic causes of lining failure, that is, the breakdown of firebrick by carbon deposition. Carbon spots are still found, especially on mortar joints, and these may be critical in the early operating stages owing to excessive expansion in the structure.

### The hearth

In the USA the bottom of the hearth is most frequently made of carbon blocks and these materials are now gaining widespread use. Effective cooling of these hearths is essential. The central hearth plug is laid with firebrick. More recently sillimanite has been used. Various combinations of carbon and firebrick are being employed throughout the world but some Russian operators, among others, have warned against the use of carbon blocks in the upper part of the hearth because they may be washed out after the blast. Furthermore, when low carbon irons are being smelted the carbon from the refractory may be dissolved in the iron, leaving the blocks more porous and making them prone to faster wear. Ramming chrome–magnesite–carbon mixtures in the lining sometimes helps to prolong the hearth life.

The deposition of carbon (*see* below) can also play a part in the life of the hearth since it may occur in fine grained carbon blocks placed in the wall of the furnace hearth. A possible cure is to use low-permeability carbon blocks. Much depends on the origin of the carbon used to manufacture the blocks (*e.g.* petroleum coke, anthracite).

Important mineralogical and chemical changes take place in hearth refractories during service. Bottom materials become impregnated with iron, carbides and graphite, densifying them and increasing their thermal conductivity. This improves the cooling efficiency and retards lining wear. The central part of the hearth bottom during operation is replaced by slag and iron, regardless of the type of refractory employed. Parnham (1966) has described the composition and some unexpected features of a blast furnace hearth after premature shutdown. This particular hearth contained spherical metal particles up to 3 in. in diameter. Near the hearth wall could be seen half-inch thick pads of a white mineral, later identified as cristobalite. The cristobalite is thought to have come from the breakdown of the hearth refractories.

## The bosh

Refractories used in laying the bosh are worn out due to the corrosion and erosion of slag attack at temperatures as high as 1 500°C and by the attritive action of lumps of burden. Furnace operation is important especially with respect to tuyere velocities; low velocities are believed by some British iron makers to affect bosh life adversely. Keen arguments continue over the best type of material for boshes and many iron makers change their minds frequently.

Carbon blocks, firebrick and even fused alumina refractories (including zirconia–alumina) are used in bosh linings. Carbon placed near the tuyere coolers are rapidly oxidised by steam, carbon dioxide or air. Tightly laid, impermeable carbon is recommended to prevent carbon deposition and carbon monoxide disintegration. Other desirable properties of bosh carbon are high thermal conductivity, high alkali resistance, high oxidation resistance and high compressive strength. The effect of potassium salts which penetrate the pores of the lining from the furnace gases and burden is important, for potassium ions expand the particles of carbon by lattice penetration and cause the brick to disintegrate.

The degree of chemical corrosion in bosh linings depends directly on the basicity of the slag and its viscosity: slags containing about 45% calcium oxide are highly corrosive. Bosh life depends on the formation of a skin containing alumina and silica from the grog (chamotte) of the refractory (for aluminosilicate type linings), enriched with lime, magnesium and other oxides from the burden.

## The shaft (stack or inwall)

The physical chemistry of stack refractories service in blast furnaces is perhaps the most interesting aspect of refractories technology as it affects the iron smelter. The classic carbon-disintegration phenomenon is only one cause of the failure of refractories in the shaft and although, as mentioned above, the use of denser kaolinitic linings has reduced the seriousness of this problem, it still exists and is more than of academic interest. In any case, there are plenty of other problems for the producer of refractories.

Pitak *et al.* (1968), comprehensively reviewing the service of refractories in blast furnace shafts with special reference to Soviet practice, sums up Russian experience as follows:

The failure of the shaft lining depends on the properties of the refractories; the system of cooling the lining; the furnace operating cycle; the action of alkalis; primary slags; zincite; deposited carbon; the furnace blow; thermal stresses in the structure; abrasion by the burden, and other factors. In order to improve lining resistance it is necessary to use dense, high-fired brick made from high-grog kaolin materials with the minimum content of fluxes; to introduce a more effective cooling system for the shafts, and to stabilise the furnace cycle, particularly at the start of the cycle after blowing and before the final drying out of the lining.

Presenting some aspects of British blast furnace practice, Brooks (1968) in an article full of practical advice on what improvements are needed in blast furnace refractories, emphasises the difference between the top and bottom halves of the stack. Sillimanite and kaolin bricks serve admirably in the top and it might be possible to leave them in place when the less satisfactory bottom half of the stack has to be replaced. A range of refractories has been tried in the bottom of the stack, including 42–95 % alumina materials (sillimanite, fusion-cast and firebrick).

Experience with lining life in blast furnaces in the USA has been discussed by Snow (1969). His conclusions, based on many years experience at the United States Steel Corporation, place emphasis on the mineralogical changes taking place during service and the importance of burden preparation, that is, the particle size of the coke, ore and limestone. British pronouncements about the diminishing importance of carbon disintegration are confirmed by Snow's findings. The problem now seems to be concentrated on mortar joints.

**Alkali attack in shafts**
In the lower half of the stack alkalis may penetrate very deeply into the lining. Alkali concentrations of 9–10 % are common and sometimes reach 25 %. In the USA damage to shaft linings has been reduced by inserting cooling plates in the structure. In the past firebrick damage was thought to be due to carbon disintegration when in fact, as recent investigations have shown, it was caused by alkali penetration. Snow (1969) has shown that alkalis can diffuse into firebrick before serious deposition of carbon disrupts the brick. Kalsite, $(KNa)_2O.Al_2O_3.2SiO_2$, forms in the brick, and one section examined by Snow contained 28·4 % $K_2O$ and 4·2 % $Na_2O$.

Petrographic studies of firebrick taken from shafts show an appreciable penetration of alkalis from the top to the bottom of the lining and marked chemical reaction with the refractory, leading to the formation of alkaline

TABLE 19.2

Properties of Refractories Used in the Lower Zone of Blast Furnace Shaft Linings

| Country of origin/use | Refractory | $SiO_2$ | $TiO_2$ | $Al_2O_3$ | $Fe_2O_3$ | CaO | MgO | Alkalis | PCE, °C | R.U.L. 2 kg/cm² | After-contraction, % | Apparent porosity, % | Apparent density, g/cm³ | Crushing strength, kg/cm² |
|---|---|---|---|---|---|---|---|---|---|---|---|---|---|---|
| Britain | 1 | 53·8 | — | 43·10 | 0·7 | — | — | 1·85 | 1 770 | 1 530–60 | 0·00 | 11 | 2·39 | 717 |
| USA | 2 | 52·16 | 2·06 | 39·38 | 3·23 | 0·35 | 0·04 | — | 1 720 | 1 408 | 0·49 | 13·3 | 2·12 | 571 |
| USA | 3 | 52·18 | 1·92 | 41·82 | 1·82 | 0·40 | 0·58 | — | 1 735 | 1 470 | 0·20 | 8·9 | 2·34 | 430 |
| W. Germany | 4 | 54·20 | 0·88 | 41·16 | 2·06 | 0·42 | 0·85 | — | 1 720 | 1 360 | — | 14·9 | 2·16 | 445 |
| Japan | 5 | 54·16 | 1·86 | 42·24 | 1·18 | 0·19 | 0·17 | — | 1 755 | 1 415 | 0·08 | 14·8 | 2·28 | 400 |
| Poland | 6 | 32·40 | 0·87 | 63·31 | 1·06 | 0·22 | — | — | — | — | — | 25·1 | — | 593 |
| Poland | 7 | 22·54 | 0·52 | 74·65 | 1·50 | 0·35 | — | — | — | — | — | 22·4 | — | 524 |
| Poland | 8 | 37–39 | 1–1·8 | 60–61 | 0·3–0·6 | — | — | — | — | — | — | 14–15 | — | 700 |
| USSR | 9 | 54·83 | 1·04 | 39·3 | 1·25 | 0·64 | 0·69 | 1·15 | 1 710 | 1 460 | 0·08 | 15·0 | 2·19 | 668 |
| USSR | 10 | 55·0 | 1·5 | 40·9 | 1·30 | 0·92 | 0·53 | 1·0 | 1 730 | 1 405 | 0·17 | 14·8 | 2·21 | 520 |
| USSR | 11 | 32·80 | 1·39 | 63·49 | 1·12 | 0·45 | 0·10 | 0·74 | 1 820 | 1 500 | 0·17 | 13·4 | 2·47 | 745 |

aluminosilicates. The structure of the refractory, especially its degree of permeability to alkalis, may be far more important than its chemical composition. Potassium carbonates and potassium cyanide, as well as the potassium oxides, may be responsible for fluxing and disintegrating the linings (*see* Table 19.2).

Mackenzie (1966) suggests that when alkalis react with low-alumina firebrick a glassy liquid forms, whereas with high-alumina material a dry powdery reaction product is formed, which may cause disruption of the refractory. A material containing 42% alumina is a 'happy medium' and can absorb alkali without growth and produce little or no liquid phase.

### Action of CO on stack linings

Because of the nature of the metallurgical process inside a blast furnace the atmosphere contains plenty of carbon monoxide. In the presence of iron oxide, as a clay impurity in firebricks, carbon monoxide is decomposed and deposits minute carbon particles in the pores of the brick. This reaction takes place at 300–600°C in micropores less than $10^{-5}$ cm in size, so the precise porosity values of firebrick made for blast furnace shafts are important. The carbon produced from the breakdown of the carbon monoxide molecule forms cementite, $Fe_3C$, with the iron contaminant of the brick, a process which is accompanied by a marked increase in volume. The brick becomes brittle and eventually breaks up, causing failure of the lining. The reaction is

$$2CO \rightarrow CO_2 + C$$

Penetration of carbon monoxide, hence deposition of carbon, is directly dependent on the gas permeability of the lining and the pressure of the gases in the furnace. As carbon particles are thought to embrittle the structure, anything reducing gas permeability will help to prevent carbon disintegration. Some operators, therefore, believe that alkali penetration, by reducing permeability, to some extent helps to reduce carbon disintegration (*see* previous section on alkalis). The catalytic action of iron oxides in firebrick can be reduced by firing the refractories at higher temperatures: up to 1 500°C, in order to convert the iron compounds into irreducible silicates. The practice today is to choose low-iron kaolin clays and fire them to a high density. The low spalling resistance may then become a problem and as in most aspects of choosing refractories, the final decision is one of compromise.

### Effect of zinc oxide

It has been suggested that zinc and zinc oxide entering shaft linings from the iron ores can catalyse the reaction of carbon monoxide decomposition mentioned above. Other workers deny this. However, zinc compounds play their own damaging role in blast furnace linings, although little is known about the way in which they are carried and deposited in a blast furnace lining. Large deposits of metallic zinc have been found in demolished stack linings (up to 6 in thick). The probable reactions are

1. $Zn + CO_2 \rightarrow ZnO + CO$
2. $Zn + CO \underset{\rightarrow}{\leftarrow} ZnO + C$

Here also, carbon is deposited and this may cause embrittlement of the structure (*see* Fig. 19.2).

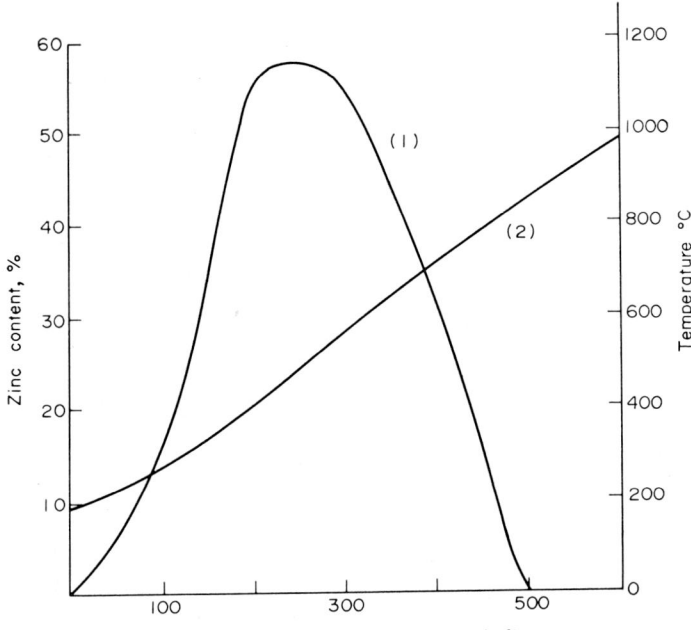

Distance from outside of blast furnace shaft, mm

FIG. 19.2.    Zinc content (1) in the shaft of a blast furnace versus lining temperature (2) and the distance from outside shaft (abscissa).

**Blast furnace stoves**
These are vertical refractory-lined cylinders filled with checkered brickwork to provide a large surface area for the absorption and emission of heat. The air blast for the furnace is preheated to 580–850°C in the stoves. The quality of the refractories used is important, since the brickwork temperature reaches 1 500°C. Blast furnace gas is burnt in the stoves for a time and then the burning is discontinued. The air for the blast is then passed through in the opposite direction to heat it up.

Water vapour and oxygen are sometimes added to the air to improve furnace performance and to keep the blast conditions uniform. Stoves last for 5–10 years. The refractories must be highly thermal-shock resistant because of the frequent changes in temperature. A firebrick with at least 45% alumina is commonly used in the high-temperature zones where the refractory usually fails because of a reduction in the refractoriness owing to the slagging action of the fluxes from the furnace gases.

Used blast furnace stove bricks are seen to have a zoned structure, the refractoriness of each zone being 100–200°C lower than that of the unused materials. This has been attributed to the formation of short prismatic mullite in the brick which, along with other furnace conditions, causes more complete separation of the glassy phase at moderate temperatures and the transformation of the material into a porcelainous structure which fuses more readily. Under the action of dust fluxes a more fusible slag skin is formed in this zone.

A range of firebricks is used to build stove checkers. In the upper part of the stoves damage occurs in the form of marked shrinkage and slumping. Slag droplets form on the dome of the stoves, eventually falling on the checkers and causing impact damage and slagging. Firebricks used in the most critical upper zones of blast furnaces last 3–4 years, in the lower zones 10–15 years and in the stoves 10–16 years.

## STEEL MAKING

Steel is made by refining and modifying the iron produced in the blast furnace described in the last section. Today a large number of different types of furnaces, involving the application of various physicochemical principles, are used for steel making. It is not proposed here to attempt a detailed description of the design of steel furnaces, since technical literature contains numerous specialist articles and books on the individual furnaces, many papers dealing specifically with the various types of refractories employed in building them. Selective references at the end of the chapter together with the papers referred to in the text should indicate where further detailed information can be found.

What is attempted in this section is an outline of the main types of refractories (although it must be said that the iron and steel industry is interested in and uses nearly all types of refractory produced) and the kinds of conditions which are common when steel is melted in contact with refractories.

When steel is being produced from pig iron during its oxidative heat treatment, slags are produced and the chemical nature of the slag, acid or basic, decides the name of the steel making process. If the slag is mainly siliceous the process is called the acid steel process, and if basic (high in lime, for example) it is called the basic process (Figs. 19.3–19.5).

The history of the steel industry can be very roughly divided into three broad periods: the bessemer steel period which lasted up to about 1910, the open-hearth period, which has persisted until the time of writing, and the period dating from the middle of the 1960s during which the basic oxygen furnace has made the open-hearth method obsolete. Developments in steel making are such that even the revolutionary basic oxygen furnace (BOF) is theoretically being challenged by the spray process. However, until the spray process finds commercial application the basic oxygen furnace will continue to produce most of the world's steel output.

FIG. 19.3. Rotating slag furnaces for studying the action of slags of various compositions on basic and acid linings. *Photo:* Courtesy British Ceramic Research Association.

FIG. 19.4.   This shows how the panel for slag testing is built up inside the furnace.
*Photo:* Courtesy British Ceramic Research Association.

FIG. 19.5.   The slag eats away the specimens to different extents as shown here. *Photo:*
Courtesy British Ceramic Research Association.

The main furnaces, therefore, still operating in steel production are open hearth, electric furnaces, and converters. The types of refractories and the reactions occurring between steel, slag and refractory materials in these installations will now be discussed.

## OPEN-HEARTH FURNACES

There are two main types of open-hearth furnace design: (a) fixed; (b) tilting. The fixed furnace is most common and is usually heated with gas and oil.

Acid open hearths are used on a limited scale for making high-grade steels by the silica reduction process which can be done only with acid (silica) refractories. However, acid linings cannot be used to remove phosphorus and sulphur from the steel and so the materials fed into the furnace must be of high quality.

The open-hearth furnace is based on the oxidation of impurities in the pig iron using a shallow pool of molten metal over which the fuel is burnt. The fuel is burnt with air which has been preheated by sucking cold air through checker brickwork (regenerator chambers) which carry a store of heat deposited in them when the furnace is reversed. It is usual to work the generators in pairs, one for air and one for gas preheating. Gas and air pass alternately into the pair of chambers every 15–20 min.

Since the temperatures of the regenerators may vary widely, from cold to 1 200°C or above, it is obvious that the materials of which they are built must be highly spalling-resistant and possess certain other refractory properties.

In the fixed open-hearth furnace the refractories have to withstand very high temperatures and sudden variations in the different sections of the structure, especially when the furnace cycle is reversed. Oxygen injection into the furnace as a means of accelerating the cycle produces great strains on the refractories because of the high temperatures that are produced, not to mention excessive dust formation and the carriage of iron oxides and batch onto the linings, causing slagging and glass attack (Fig. 19.6). Slagging may be particularly serious in the roof of the furnace, and of course in the checkers since the volume of dust-laden gases passing through the brickwork is very high.

In the tilting open-hearth furnace provision is made for the working hearth to be inclined by 15–35°C. Slags can thus be poured off when necessary and it is possible to cast part or all of the steel in one operation. The main difference between refractory service in fixed and tilting open hearths is the more adverse conditions in the roof and walls of the bath in the latter. Obviously when molten steel is being swirled around the furnace the effect on the refractory lining can be quite devastating. In

FIG. 19.6.   Effect of excessive temperature on two different grades of firebrick in a
side wall of a steel furnace below a suspended roof. *Photo:* Courtesy M. H. Detrick.

TABLE 19.3

*Open Hearth Steel Output, % of Total*

|            | 1964 | 1966 | 1968 |
|------------|------|------|------|
| Britain    | 70·0 | 59·0 | 54·8 |
| USA        | 77·3 | 63·4 | 50·1 |
| W. Germany | 45·0 | 39·0 | 35·3 |
| France     | 26·0 | 22·9 | 20·0 |
| Japan      | 34·8 | 18·1 | 8·1  |

extreme cases when the furnace is being tilted the roof may get twisted and so special supports are built into the design. Therefore, it is not surprising that the life of tilting furnaces may be only half those of fixed furnaces.

Table 19.3 shows data illustrating the decline of the open-hearth furnace in the world steel industry over the period 1964–1968.

In 1969 the new basic oxygen furnace displaced the open-hearth furnace as the chief unit in the steel industry of the USA. In the last five years no new open-hearth furnaces have been constructed in the USA or Great Britain.

**Open-hearth roofs**
Silica (dinas) was once the only roof making refractory for open-hearth furnaces owing to its high refractoriness under load and constant dimensions at high temperatures. Today magnesite–chromite and magnesite have replaced silica in most roofs, mainly because, with higher temperature melting resulting from the use of oxygen, the fusion characteristics of silica are inadequate.

The design of magnesite–chromite roofs is far more complex than that of silica roofs which are extremely simple in design, but the increased costs and trouble are usually recouped in the longer lives and higher productivity.

Fig. 19.7.    Fully suspended hold-up, hold-down basic refractory roof of an open-hearth furnace before the start of the campaign. *Photo:* Courtesy M. H. Detrick.

The drive for better roofs and the need for more highly resistant linings in general has led to a greater output of magnesia (especially from sea water). The proportion of magnesite and magnesite–chromite refractories in the total output of the world's refractories industry has risen rapidly: in the USA from about 50% in 1960 to more than 83% in 1970; in Britain from 30% in 1960 to 60% in 1969, and in Japan from 70% in 1960 to 82% in 1969.

FIG. 19.8.    Open-hearth roof after campaign (see Fig. 19.7). The used roof has been cut away to expose the method of installation. *Photo:* Courtesy M. H. Detrick.

At the time of writing nearly all basic open-hearth furnaces in the USSR are built with roofs constructed of magnesite–chromite brick.

Improvements in roof refractories have been achieved by using purer raw materials, reducing the porosity and increasing the hot strength. Roof design is also important (Figs. 19.7 and 19.8).

Currently, raw materials such as magnesite and beneficiated chromite are being calcined together in rotary kilns at 1 900°C to yield super-quality magnesite–chromite in which many of the essential reactions of direct bonding have occurred in the clinker product. The clinker is suitable for making unfired chemically bonded bricks. These roofs have lasted for about 200 heats.

Neely *et al.* (1970) have described sintered magnesia–chrome grains for new types of refractories, which have been tried in electric-arc furnace sidewalls. This new type of refractory raw material, which is essentially a 60% magnesia–chrome grain, shows the microstructural features and physical characteristics of the direct-bonded magnesia refractory. The developers claim that it is homogeneous, has a low and uniformly distributed $SiO_2$ concentration, a low apparent porosity and various other advantages for steel plant construction.

Herron and Smothers (1969) have described the production of fine-grained magnesite–chrome refractories, the strength, spalling resistance and slag resistance of which are said to be much better than those of direct-bonded and fusion-cast magnesite–chrome refractories. The authors credit the improvement in properties of the fine-grained material to the retention of a bigger volume of closed porosity, following high-temperature heat treatment. The development of the closed pores is controlled by reactions between the fine material and the quantity and distribution of the silicate phase. The importance of raw material purity in developing direct-bonded and new types of refractories for steel making is emphasised by these workers in their reference to the great sensitivity to the silica content of the raw materials used for making magnesite–chrome refractories. They state that an $SiO_2$ concentration of 2·3–2·7% is optimum to obtain the minimum open porosity and sound refractoriness-under-load values.

In some countries magnesite is being calcined at 2 200°C to give more stable roof refractories.

Direct-bonded bricks (magnesite) have been widely mooted recently for open-hearth roofs but the cost may be 30–50% higher than ordinary magnesite, and there seems to be no great increase in campaign life from direct bonding.

The capacity for stress accommodation in open-hearth roofs is one of the most important operating factors and it is possible that the use of high-density, stable refractories such as direct-bonded materials may not provide this accommodation. These remarks may also apply to other elements of furnace design, notably in electric-arc furnaces. It may be better to use unfired or low fired refractories. When metal plates are used in the roof structure of open hearths for bonding there is also less chance of their assimilation by the refractory in order to yield the desired monolithic structure when the refractory bricks are excessively dense and tight. The roof may then be found to expand by up to 3% which increases the stresses. A possible solution to this problem is to use thin packing made of corrugated plates, asbestos, or cartonboard, or to use clad bricks.

Fusion-cast refractories are now used in open-hearth roofs and will probably gain favour in the steel industry for other applications since costs are reduced with the development of this method of fabrication.

Guncreting is another new technique now being applied to open-hearth formation throughout the world in places where these furnaces are still operating. For example, one roof working without oxygen and built of unfired magnesite brick had its campaign doubled to 800 heats by guncreting. The refractory concrete is applied after 6–9 heats from the commencement of the campaign and then every day, thus eliminating the need for hot repairs.

### Breakdown mechanism of open-hearth roofs

The reactions going on in an open-hearth roof are very complex but it may be useful here to outline the chief reactions. Iron oxides are taken up by the magnesite–chromite roof to form magnesioferrite and solid solutions with periclase. Chrome-spinels are formed which also take up iron oxides to develop other solid solutions with a large volume expansion, leading to bursting of the chromite grains in the refractory roof. The consequential stresses, if not accommodated, as mentioned above, may cause spalling, cracking and flaking. Reducing the formation of zones in the roof is seen as one of the main techniques of improving its life, and this is done by increasing the density of the brick and seeking to achieve the optimum chemical composition.

Under reducing conditions the melting point of iron–magnesia refractories falls rapidly and care must be taken to ensure that the fuel is properly combusted.

Magnesia-spinel brick with a low porosity has been recommended for basic open-hearth furnaces.

### The hearth

The hearth of the basic open-hearth furnace consists of a lining of magnesite brick on top of which is laid a layer of magnesite and other basic powder. Firebricks, insulating bricks and asbestos may also be used as backing layers. Dolomite bricks and fettling materials are also used. Since the hearth bottom is to hold the molten steel it must be fabricated carefully to withstand the metal penetration and breakout which is likely. Chesters (1963) has described the construction of different types of hearth including rammed hearths which have replaced fritted hearths.

During early melting the hearth temperature may reach 1 600°C, and during the final stages it goes as high as 1 750°C.

Fettling is the process of repairing or remaking open-hearth linings (and other furnaces) by throwing in quantities of powder while the furnace is still hot. The powder, consisting of mixtures of various refractory materials such as magnesite and dolomite and iron scale, formulated to optimum grain-size distributions in order to stay in place, is consolidated and sintered by the heat of the furnace to fabricate a monolithic lining. Fettling goes on throughout the campaign of the furnace. The fettle layer is made up of magnesiowustite cemented in a consolidated mass by calcium ferrites and silicates.

Acid open-hearth bottoms are made from quartz materials by sintering and consolidating layers of siliceous powders such as quartz sands. They wear out more rapidly than basic hearth bottoms.

**Other design elements**

The other critical parts of open-hearth furnaces include the front and back walls and the checkers. The front wall located near the charging end and made up of several columns of brickwork has to withstand batch dust, slag and metal splashing, as well as physical impacts from the charge, so it does not last long and needs frequent replacement.

Chrome–magnesite and metal-clad magnesite–chromite have now largely replaced silica as the main refractories in the front and back walls. The back wall usually slopes by as much as 45°. It contains apertures for discharging slag and steel. Usually the outside is well insulated with firebrick and coated on the inside with a layer of fettled magnesite. In service the back wall has slightly less harsh treatment than the front wall, mainly because it does not have to stand the impact of the metallurgical charge.

*Checker chambers*

Combustion products from the open hearth pass through checkers at temperatures of about 1 500°C at a speed of up to 0·5 m/sec. They leave at temperatures of 500–600°C. The checker's upper structure is made of silica brick with the roof being built of chrome–magnesite or high-alumina refractories. Firebrick of various qualities is used in the lower sections. Gas tightness is an important factor in checker refractories.

Fireclay, silica, high-alumina, and basic refractories of various compositions are used for checker making. Since the task of the brickwork in checkers is that of heat exchange, and since the complex network of surfaces has to be kept in the optimum heat-receiving and heat-emitting state it is necessary to ensure that the channels do not get clogged up with dust and deposits. Silica has a high heat-exchange capacity but under the action of ferruginous dust the brickwork may get clogged up with iron slag. At temperatures above 1 200°C slagged silica checkers soon fuse, causing the checkers to fail in service.

Firebrick checkers may also be affected by the build-up of skin or scum from the furnace dust, hindering efficient heat exchange. The drive for higher operating temperatures, the use of oxygen, and basic roofs have all combined to shorten the life of checker bricks.

Forsterite refractories are sometimes used in air checkers (fired and unfired). They are laid on top of firebrick, not silica, because of the danger of reaction between the two materials which weakens the structure. Other basic bricks used for checkers are of magnesite (fired and unfired) which were tried in the steel industry, following successful trials in glass furnaces, and also magnesite–chromite refractories. Studies have been made with the use of chrome–forsterite brick in open-hearth regenerators and in the slagging chambers.

The design, operation and materials technology of open-hearth checkers vary widely in different countries of the world and it is misleading to generalise about the refractories used and the temperatures prevailing. For example, some steel producers lay great store by cleaning and blowing procedures, while others claim that these have little effect. The properties of the burden, the temperatures, the use of oxygen or not, and the design of the checkers all affect the operating procedures and the life of the checkers.

## ELECTRIC STEEL FURNACES

Furnaces in which the source of heat for smelting steel is electricity are of two main types:

(a) electric-arc in which the arc is struck between electrodes (graphite or amorphous carbon refractories) near molten metal or scrap; and

(b) induction, in which an alternating current is passed through a coil surrounding the crucible containing the charge to be smelted. Currents are induced in the metal charge and the resulting heat smelts the metal.

As in the open-hearth furnaces described above electric-arc furnaces are either acid or basic depending on the refractory linings of the furnace housing. A typical 80–90 ton electric-arc furnace might have a magnesite lining backed up with fireclay and insulating refractories. Some designers may prefer to use unfired clad magnesite–chromite; the bottom may be made either of firebrick or ramming powders.

Small acid-arc furnaces (0·5–5 tons) frequently have linings (walls) rammed with siliceous refractory compositions. Replacing the brick structures by ramming materials has increased the life of some furnaces from 20–80 to several thousand heats, according to Soviet research. The practice has not become widespread, however, since it is usually necessary to hot fettle after every heat and working conditions are very adverse.

Since they do not come into contact with molten steel the roofs of basic arc furnaces have usually been made of high-quality silica brick; for smaller furnaces sillimanite or other high-alumina products may be used. In Britain at least, early trials with magnesite, chrome–magnesite and forsterite were considered not to have warranted the high cost, except for smelting special alloys. However, by 1961, 95% of all electric steel furnaces operating in the USSR steel industry had basic roofs. This lengthened their life, enabled oxygen to be used in order to speed up melting, and increased the voltage and capacity of the furnaces.

According to Bakker and Snyder (1971) silica brick in electric furnace roofs has been largely replaced by 70% alumina brick, while superduty firebrick in hot metal cars is being replaced by 60% and even 85% alumina brick. These authors also state that large induction furnaces used for melting or holding iron in foundries are frequently lined with high-purity 90% alumina brick.

The walls of basic electric-arc furnaces are lined with various materials, including mixtures of magnesite and burnt dolomite, and magnesite alone. Iron rods 25–30 mm thick are sometimes used as reinforcers in some furnaces. The materials may be placed as blocks of compacted powder, or as loose powder.

Other refractories employed are magnesite–chromite brick and unfired clad magnesite–chromite. Guncreting and ramming with plastic refractories are useful techniques for small furnaces. For instance, a composition containing 88–94% magnesite 0·3 mm, and 12–16% fireclay mixed with water glass has been successfully used.

Acid-arc furnaces can be lined with quartz sand bonded with 10–12% water glass.

Sarjant (1958) has described the principles and design of electric melting furnaces and outlined the types of refractories needed for the various furnace sections, although it must be stated that developments are taking place very rapidly in this field and the types of refractories and methods of use are currently undergoing considerable change.

### Service of refractories in electric furnaces

During operation the rammed bottom of an electric-arc furnace becomes saturated with silica, calcium oxide and MnO. This impairs the refractoriness, and repairs are usually necessary using fine magnesite powder after every heat. Some operators use mixtures of magnesite and dolomite for the repair job, while others are employing magnesite powder mixed with chromite in various proportions for the smelting of stainless steels. Russian operators have experimented with a ramming body made from high-fired magnesite, containing no free lime, bonded with sulphite lye.

The walls can be made in different ways; for example, rammed powder blocks as mentioned above and ordinary bricks. The rate of wear in the walls of arc furnaces varies from top to bottom. Lower sections fabricated, for example, from magnesite–chromite brick wear out 3–4 times more quickly than the upper sections. Linings made from blocks may wear out even more rapidly.

In electric induction furnaces the walls must be as thin as possible because thick walls prevent the contact of the electric field with the burden; no possibility of short circuiting should occur.

## CONVERTERS

As mentioned above the basic oxygen furnace (BOF) is currently rapidly replacing the open hearth throughout the world's steel industry. However, until now the best known converter used by the steel industry has probably been the Bessemer converter. From 1910 onwards, the Bessemer had to take second place to open hearths, but now the converter process has won back most of the lost ground in the form of the basic oxygen furnace, of which there are several versions.

Described simply, a converter produces steel by the process of blowing air or oxygen through molten iron which contains scrap and fluxing agents. Many basic oxygen vessels in use are modified Bessemer converters in which the oxygen is admitted through a lance (a metal pipe) inserted from above instead of through the bottom. The obsolete tuyeres of the Bessemer are rammed solid. The main advantage of converter production is speed of operation.

In oxygen converters the vessel can be tilted to pour out the molten steel. During the passage of oxygen (about 2 tons of oxygen are needed for each ton of steel) through the metal, the phosphorus, carbon and other elements are oxidised, with the release of exothermic heat; thus no additional fuel or heat input is required. Such conditions—high temperatures, slag and metal erosion and the swirling action of the metal during pouring—make for very severe operation, and satisfactory refractory linings largely determine the success or failure of the operation.

However, the demise of the open hearth and the rise of the basic oxygen converter in the steel industry means that there will be considerably less demand for refractories of all types in the steel industry. This trend will also be strengthened by the development of continuous casting and other new steel-processing techniques (*see* below).

During the early development period of the Bessemer converter when the iron was frequently smelted from low-phosphorus ore the converter lining was made of firebrick or other siliceous material, using clay mortars (acid converters). Later, basic converters were used in which the linings were made from lime refractories. Sydney Thomas, a pioneer in this field, used a refractory lining made of burnt dolomite and coal-tar pitch with which the metal could be rapidly dephosphorised and the wear of the lining retarded.

The arrival of cheap tonnage oxygen has meant a great resurgence of the converter method in steel making.

The oxygen method has necessitated the development of new types of refractories. The chief one has been the tar-bonded dolomite refractory. Other refractories of interest to the user of basic oxygen converters at the time of writing are tar-bonded dolomite–magnesite, and tarred magnesite.

## Acid converters

A typical acid converter of the Bessemer type has a lining made of silica brick with a quartz clay rammed bottom and with the tuyere section built of firebrick. The molten metal has a temperature of 1 700–1 750°C in the hottest part during the hot time of the blow. The lining, usually made of soft-fired dinas, is destroyed as a result of mechanical attack from the charge and the corrosive action of a double silicate of iron and manganese (the slag). The bottom of the converter, coming under the most severe attack, usually fails first near the tuyeres, because the jets of air (oxygen) entering through the tuyeres vigorously oxidise the iron impurities, with the generation of very high local temperatures. The bottoms of acid converters do not usually last more than 40–50 heats.

## Basic converters

These are more important to steel making than acid converters. Basic oxygen converters with capacities of 60–200 tons of steel are now in common use. A few furnaces have capacities up to 300 tons and higher. Since the blast of oxygen near the base of the converter would cause the relatively weak tuyere region to fail prematurely, the oxygen is fed in from the top. The process is known as oxygen lancing.

The lining of basic oxygen converters is done in many different ways with many different refractories. Steel making in converters occurs very rapidly and vigorously. Ten minutes after the start of oxygen blowing, the steel temperature has risen to about 1 570°C but the reaction temperature (when the iron impurities are being oxidised) may go as high as 2 500°C. The most severe conditions are experienced at the site where the oxygen jets contact the liquid metal.

### Causes of wear

The principal factors of lining wear in oxygen converters are the mechanical damage from batch materials and scrap, the blow, the high temperatures, and the slagging cycle. The hydrodynamic action of the currents of slag and steel and gaseous erosion are other factors.

According to Kuznetsov et al. (1970), of these factors, the wear of tar-bonded dolomite–magnesite linings is affected mostly by the content of iron oxides in the slag and the temperature. When the oxygen blow-time is reduced the lining lasts much longer. Increasing the content of calcium oxide in the slag, thus raising its basicity, also reduces lining wear (Fig. 19.9). The rate of solution of the lime during the blow is not constant. Kuznetsov recommends that a faster increase in calcium oxide concentration in the slag is helped by making a second charge of limestone before it begins to dissolve intensely, that is, 4–6 min after the start of the blow. The use of multinozzle lances also gives slower and more uniform lining wear.

Nepsha (1967) discussing Soviet research up to 1966 with converter tar-dolomite refractories, outlines the basic factors in the operation of these materials. No water should be allowed to come into contact with the linings prior to use. The downtime between heats in which air gets to and oxidises the tar should be reduced. Throwing in slag-metal batch material at high temperatures (up to 1 550°C) causes serious damage to the lining, the compressive strength of which is only about 2 kg/cm² at these temperatures. According to Nepsha, lining life depends largely on the concentrations of silica, phosphorus, alumina and iron oxides in the slag, since these readily react with the calcium oxide and magnesia of the lining to form fusible compounds which enter the slag.

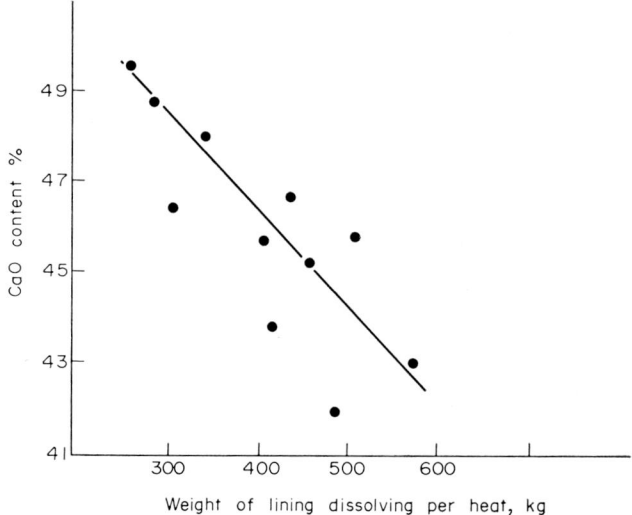

FIG. 19.9   Effect of CaO content in slag on lining wear in a basic oxygen converter. After Kuznetsov *et al.* (1970).

Three visible zones are formed in the tar-dolomite lining: (1) flux zone; (2) decarbonised; (3) carbonised. The carbonised zone is impregnated with a residue of coke; it also contains calcium and magnesium oxides, together with calcium silicates acting as bond, metallic iron, and an amorphous carbon substance. Graphitic carbon may also be present. The resistance of carbon and graphite to molten slags and metals is well known. Thus, the carbon zone is completely untouched by the slag which penetrates from the flux or slag belt to the decarbonisation zone.

The flux zone, dark brown in colour, is composed of magnesia, $3CaO.SiO_2$ and $2CaO.SiO_2$, together with a liquid phase. Towards the hot face the amounts of refractory phase diminish and the concentration of liquid phase may reach 50%.

The reactions between slag, dolomite and tar are very complex and not fully understood, but furnace operators now know that any move to successfully control these reactions (by modifying the lining composition, matching the slag cycle, etc.) can reduce lining wear.

## POURING PIT REFRACTORIES

After the steel has been smelted and refined and is ready for discharge from the furnace it is cast or poured into a variety of moulds or containers. The handling of molten metal at high temperatures needs the use of a range of highly reliable refractory utensils, including ladles, stoppers, nozzles, sleeves, moulds, hot tops and casting assemblies. That is, all those articles used to transfer the steel from the smelting furnace to the moulds.

The properties of these refractories are not easy to specify on the basis of generalisations because steel-pouring practice varies so much from plant to plant and country to country. Some applications call for highly resistant materials which can be used repeatedly, whereas other processes involve using a low-fusing point material once only. Sometimes the refractory is 'absorbed' by the steel being poured and produces instant slag which may rise to the top of the molten metal. It can be quickly removed and does not harm the quality of the steel.

Continuous casting, which has developed rapidly in the last few years, is eliminating the need for some casting refractories such as mould tops and guide pipes but it has raised other problems for the refractories maker.

The extensive use of oxygen in steel making has led to the pouring of metal that has a much higher temperature and this too has increased the responsibility of casting refractories.

Vacuum casting of steel is a further new development tending to alter casting conditions and demanding new types of refractories both in shape and physicochemical properties.

All these new developments are going hand in hand with research to discover how refractories behave in service. Naturally, since the steel maker has taken a lot of trouble to refine his product and get it in the desired chemical and physical state, any subsequent contamination by non-metallic inclusions from the refractory utensils employed for casting it from the furnace will be most undesirable. New techniques of studying the influence of refractories on steel quality and the part played by casting pit refractories in contamination by these inclusions, for instance, using radioactive isotopes, have been developed and are now widely used.

The vast majority of casting pit refractories are made with alumino-silicate raw materials, chiefly siliceous and aluminous fireclays. Experiments have been carried out with basic refractories and with combinations such as graphite and clay (which are still in industrial use). In vacuum casting where slag suction into the brick may be severe, chrome–magnesite brick has been successfully used. Special types of alumina refractories in which the bond of the grog–clay material is impregnated with high concentrations of alumina, altering the phase composition of the bond, are also of special interest.

TABLE 19.4

*Temperatures of Steel During Pouring*

| Steel Type | Tempera-ture, °C |
|---|---|
| Killed open hearth | 1 540–1 590 |
| Rimming open hearth | 1 570–1 600 |
| Electric | 1 600–1 670 |
| Bessemer converter | 1 600–1 640 |
| Oxygen converter | 1 610–1 660 |

As soon as casting begins the temperature falls by 50–60°C, and by completion it has fallen by 90–100°C.

**Ladle bricks**

The thickness of ladle linings varies with the size of the ladle and type of steel, but all ladle bricks have one common essential property: they must be able to contain the molten metal in the ladle. This has usually been interpreted in terms of high density and high chemical resistance to molten metal and slags. However, the sudden splash of metal at temperatures of about 1 600°C, followed by cooling in air, means that ladle linings must be thermal-shock resistant and if they are too dense spalling may occur. Ladle tightness is often achieved by using expanding bricks such as siliceous or kyanite.

Ladle brick performance cannot be measured merely in terms of the number of heats sustained by the lining since the thickness of the lining and the size of the ladle are critical factors. A more effective measure of comparing the performance of ladle bricks is the actual amount of wear of the brick recorded in millimetres per heat. This can be done, of course, only after the ladle has been dismantled.

The wearing away of ladle linings takes place during the time the molten metal and slag remain in them. It is found that the degree of damage is not uniform from top to bottom of the lining but increases towards the bottom. However, the rate of wear during the hot time is greater at the top than at the bottom. It is usual therefore to vary the lining thickness over the height, in order to save refractories, in proportion to the degree of wear.

Basic open-hearth slags which float on top of the metal in the ladle, preventing it from being cooled and oxidised, damage the ladle lining more than does the steel. The metal remains in the ladle from 10–30 min, during which the upper rings of lining are exposed to corrosive slag attack, while the rest of the ladle is in contact with metal.

## Chemical processes

When they are first poured into the ladle, basic slags, being very hot and fluid, are very corrosive. First contact between slag and ladle leads to an increase in the silica content of the slag and a fall in basicity, but as casting proceeds the slags thicken and are less prone to reaction with the acid lining. Further, as the firebrick dissolves in the slag, and as a result of the slag–metal reactions, there is a change in the chemical composition of the deoxidation products. Any fall in the iron-oxide content and in the basicity of the slag diminishes the softening action of slag on the firebrick lining.

However, in some ladles the deoxidation process causes an increase in the manganous oxide content of the slag, and this increases its activity. Soft, killed steels, the slags of which have high MnO contents, cause very rapid corrosion of ladle linings. Any reduction in slag basicity tends to make it more fluid and more capable of penetrating the lining. The chemical composition and physical state of slags being poured into ladles is thus a matter of careful balance.

## Zone formation

The normal wear of firebrick in steel ladles occurs as a result of the formation of three or four zones, moving from the hot inside of the ladle to the cool outside. The inside is coated with a shiny dark slag layer making up the main working zone; its thickness is usually less than 1 mm but this depends on the composition of the refractories and slag. It is made up of about 40% firebrick and 60% slag with a melting point of 1 220°C, with some of the firebrick being dissolved in the slag. The second layer is a densified layer 5–10 mm thick, almost indistinguishable from the first, but it is found to be less porous and slightly lighter in colour. If the ladle wears out very rapidly the demolished lining may not have this second zone. The third zone is again dark in colour and passes into the unaffected brick.

Chemical analyses of zoned bricks taken from ladles suggest that the slag action is restricted to the working zones. The second zone receives diffused ferrous oxide, manganous oxide and magnesia, but the third zone receives only magnesia from the basic slag. The lighter colour of the second zone has been attributed to the formation of mullite crystals.

Ladle design is an important element in the life of steel ladles. The way in which the bricks are used, the quality of the mortar (the chemical composition of mortars is chosen in accordance with the refractory brick) and the method of laying are also important. For example, bricks laid with vertical joints wear out quicker than with horizontal joints.

The thickness of the slag layer has a marked effect on ladle lining wear (*see* Fig. 19.4).

Monolithic linings are used in some ladles. These are made by ramming sand–clay or grog–clay mixes.

### Ideal properties of ladle brick

Ladle brick made by plastic moulding methods sometimes contains internal cavities and when the steel penetrates through the face of the brick these cavities can represent critical sources of subsequent devastating attacks. The properties of the surface layer of a ladle lining are immediately altered upon the first contact with slag and metal. It is, therefore, the slag–brick reactions which occur that are of importance in making ladle bricks. The slag-impregnated layer is usually glassy and dense. It cannot withstand thermal shock. Very roughly, these effects have more serious consequences for the lining life the lower the grog (chamotte) concentrations, and the higher the content of plastic clay used in the brick. It is found that high-grog bricks made by pressing techniques last longer than plastic-formed products since they absorb less slag and the erosion of the grog grains is more uniform.

Another important factor in compiling grog–clay mixtures for ladle brick is the temperature at which the grog is fired. Some producers claim better results with soft-fired than with hard-fired grog.

The quality of ladle brick is of the utmost importance for steel quality, since it is from the surface of ladles that steel receives its non-metal contamination: a source of great potential loss in steel plants. Methods of improving ladle brick, either by perfecting existing techniques with existing materials, or by developing new materials, are therefore of great value.

### Silica or alumina for ladles?

Some controversy exists about the chemical composition of the most suitable ladle refractories. In Britain, the United States and other western countries, steel ladles are widely made on the basis of fireclay and other siliceous refractories, as mentioned above, and so far no evidence has been produced by research to show that these refractories should be replaced by high-alumina products. In the Soviet Union, on the other hand, advocates of the use of high-alumina ladle linings declare that there is no basis for imitating western experience. Soviet developers are, therefore, currently concentrating on producing special high-alumina and other aluminous ladle brick. Efforts are being directed at perfecting alumino-silicate refractories with high alumina concentrations and low apparent porosities. The examination of the theoretical principles of the behaviour of refractory mixtures of refractory and slag indicates that semi-acid (siliceous) refractories have no advantage over fireclay. The alumina content coming within the range 25–40% has no great effect on the refractoriness of such mixtures. According to Aristov (1969) when the mixture contains 40% slag a rise in refractoriness is noted only for products containing 55% alumina; with 50% slag concentration for refractories containing 60% alumina. Repeated experiments with high-alumina ladle brick containing these quantities of alumina have shown that they are highly resistant and wear out very slowly during steel casting, but there have been no noticeable increases in the campaign length of the ladle. The adherence of slag and growth of the ladle lining has been observed with the use of high-alumina brick containing 57–70% $Al_2O_3$ when the batch contained commercial alumina. According to some Russian workers in this field, the cause of the sticking is the presence in the brick structure of free corundum, which upon reaction with basic open-hearth slag leads to the formation of a highly refractory complex spinel in the contact layer of the lining. In the opinion of these workers, the linings of the ladles made from high-alumina brick can be very resistant if the crystal phase in the brick is mainly mullite and not corundum. This can be arranged by using kyanite; also kaolinite mixed with hydrated alumina. It is concluded that the phase composition of the ladle brick is more important than a high alumina concentration. This conclusion is very important and to some extent could prevent an increase in the use of expensive alumina brick for lining ladles.

In aluminosilicate refractories, as with other types, the weakest part of the lining is the bond between the grog grains. Over the years many attempts have been made to reduce this weakness and one technique is to incorporate calcined alumina into the body, or more particularly into the bond constituent of the batch. Slag resistance tests with brick made from ordinary chamotte and a high-alumina bond showed that they have a high resistance to slag if the bond is made from precalcined, fine-ground alumina and clay. The optimum alumina content in the bond is considered to be 55–60%, and the alumina concentration in the product as a whole should be 46–48%. Any further increase in alumina concentration does not increase the slag resistance. The boosting of slag resistance by increase of the alumina to 55% is claimed to be due to the reaction between the slag and the corundum in the bond of the refractory which, with the stated alumina concentration, is the primary crystallisation phase. With an alumina concentration of 45% the primary crystal phase is mullite. The general conclusions of Soviet research in this field have been used for perfecting a method (at the East Institute of Refractories) for making improved ladle brick. The main feature of the technique is to incorporate fine additives, obtained by the combined grinding of clay and high-alumina chamotte which are specially prepared, into the bonding constituent of the batch. The latest trials indicate that the optimum alumina content in the mix is 55–62%.

According to Ignatov *et al.* (1964), the advantage of the brick with the high-alumina bond is the fine porous structure of the bond with a predominant pore size of 0·002–0·004 mm and the presence of microcracks between the grog grains and the bond. This, apparently, is a favourable factor on thermal-shock resistance, which, with an apparent porosity for the brick of 10%, equals 11 heat cycles and with a porosity of 16·2%, 19–23 heat cycles.

### Carbon–fireclay ladle bricks

The addition of carbon materials such as graphite to clay to make ladle linings has been practised for many years. The optimum addition of graphite is about 30%, depending on the type of steel being cast. Chesters (1963), mentioning British and Russian work, pointed out the difficulties with the use of plumbago (graphite–clay) brick in connection with the development of suitable cements for bonding linings. Chesters also mentions that the mechanism of the high resistance, whether it is physical or chemical, is unknown. Subsequent research has been concerned with the relationship between slag resistance and the content of graphite in these products. More recent experiments (1965 onwards) have confirmed that the problem of providing a suitable mortar still remains. Some experiments involved using ground ferrosilicon, a step which has reduced the wear attack on the joints, but even so the campaign of the ladle terminated after about 12 heats, due to the wearing away of the vertical joints in the lining. A serious disadvantage with graphite–clay ladle brick is the resulting high thermal conductivity due to the presence of graphite, a phenomenon which affects the cooling rate of the steel in the ladle. One way of harnessing the advantage of graphite to ladle application may be to reduce the graphite to 10–15% in order to reduce the thermal conductivity.

Another form of carbon–clay composition of interest to ladle brick manufacturers is the tar-dipped firebrick. The procedure is to immerse the firebrick in the hot tar, followed by firing at temperatures up to 800–850°C, resulting in the deposition of coke in the pores of the brick and reducing the porosity by as much as 25%, depending on the degree of penetration of the tar in the brick. Ladles lined with tar-dipped brick have lasted for up to 25 heats, compared with 12 heats for untreated brick.

### Other casting pit refractories

As mentioned above, the development of continuous casting in steel practice will eliminate many of the refractory components described, for example, in British Standard 3446:1962 in the section dealing with steel casting (centre brick, runner brick, trumpets, etc.). However, the run-off channel (lander) will continue to be of importance, as will the design and production of the ladle spout and other ancillary products. From the steel-smelting furnace, such as the converter, the steel flows into the ladle through the run-off channel. The bottom and walls of the run-off channel are usually lined with standard firebrick. The most corrosive action occurs in the lower side walls near the base due to the fast flow of slag and metal. Sometimes the firebrick lining is protected with a clay–graphite–grog wash, bonded with water glass to give a temporary bond.

### Unfired material in casting pits
Recently great interest has been expressed in replacing fired brick by castables and ramming materials or by spraying coatings onto conventional firebrick in the formation of ladle linings and for lining run-off channels and other parts in steel casting pits. This technique reduces the reliance placed on skilled bricklayers and lends itself to mechanisation. The physical structure of the rammed and coated linings is of vital interest, especially in relation to slag and metal penetration. Materials with large pores are prone to rapid attack from molten metals and slags. Since the fine-pore structure of a refractory develops during the calcination, heat treatment or kiln firing, the formation of a lining cast *in situ*, with the development of a large number of fine pores, may produce better results in ladles than the use of conventional firebrick. Some of the features of the use of rammed linings and castables were discussed in Chapter 18.

### Continuous steel casting
The development of continuous steel casting will probably eliminate the use of soaking pits and other structures needing large tonnage refractories. However, the quality of the refractories which are used will be much higher and super-refractories, such as zirconia, will be required for making, for example, tundish nozzles which need to have very long lives. Briefly what happens with the continuous casting process is that the liquid steel leaves the furnace and enters a tundish where the slag is removed. The steel is then cooled in a copper crystalliser and solidifies. The steel ingot then passes into rollers where it is turned into a horizontal direction followed by cutting into appropriate lengths. The continuous process retains ladles and refractory tundishes. The ingenuity of the refractories producer is being tested in designing suitable refractories for this advanced steel-making process.

Humphreys (1969) has outlined the problems and some of the features of making refractories for use in the continuous casting of steel. He suggests that future developments are likely to be the use of bigger machine capacities *e.g.* up to 270 ton ladles (as used in the Soviet Union) and also rapid increases in the use of sequential casting, using several ladles of steel which are cast through a machine without process interruption. Tundish and teeming refractories must withstand long pouring times. The advantages of continuous casting are considerable: the production cycle is reduced and several laborious and difficult processes are eliminated. Scrap metal may be reduced to about 5% instead of the present 15–20% with ordinary casting. There is also the distinct possibility of complete automation of the steel producing cycle. Capital investment in constructing steel plant is also much less than with conventional open hearth–casting pit technology.

The main features of the ladle brick conditions and stopper refractories in continuous casting plant are the increased metal and slag temperatures,

TABLE 19.5

*Metal Temperatures for Soviet Steels Made By Continuous Casting, °C*

| Steel | In furnace before tapping | In ladle | In intermediate ladle |
|---|---|---|---|
| Killed carbon | 1 615–1 635 | 1 585–1 615 | 1 525–1 550 |
| Rimming carbon | 1 630–1 650 | 1 590–1 615 | 1 525–1 545 |
| Transformer | 1 605–1 625 | 1 590–1 610 | 1 530–1 550 |
| Dynamo | 1 620–1 640 | 1 605–1 620 | 1 540–1 555 |

owing to the need for prolonged dwell of metal in the ladle (more than two hours) and the use of intermediate ladles which is technically essential by virtue of the process. Soviet experience has been of particular interest in the development of the continuous process. Table 19.5 illustrates the temperature limits of various types of Soviet-made steel using the continuous process.

The data in Table 19.5 refer to experience at the Novolipetsk Metallurgical Factory. The life of the linings of the ladles used at this plant for killed carbon steel was 8–9 heats; for rimming carbon 9–10 heats; and for transformer steel 4–6 heats for firebrick linings and 6–7 heats for kaolinised linings.

Developments in the production of various types of refractories for continuous steel casting are still going on, and it is difficult to generalise about the properties or conditions encountered in continuous steel casting. Reference should be made to the current literature for information on this subject.

## VACUUM TREATMENT OF STEEL

The aims of the steel producer include the production of high-quality steel containing the minimum concentrations of hydrogen, nitrogen and oxygen. This can be achieved by vacuum processing the liquid steel. Three methods are possible: (1) removing gas from the steel in the ladle established in a vacuum chamber, followed by pouring the molten steel at atmospheric pressure; (2) vacuum processing by tapping the metal from an ordinary ladle into another ladle which is housed in a vacuum chamber; (3) gas removal during casting of the ingot in the mould housed in a vacuum chamber.

When the refractories are used for the vacuum processing of steel the pressure inside the pores of the refractory is reduced and the gases previously housed in the pores are removed. When the pressure is restored to atmospheric, particles of slags, deoxidation products, and oxides suspended in the steel, as well as the steel itself, may penetrate into the refractory. This is particularly true of large pores, cracks and cavities in the refractory materials as well as in the mortar joints. Thus, the wear of the refractory in the ladle and other containers and especially the stopper tubes may be greatly enhanced. The vigorous boiling of the steel under vacuum treatment also causes rapid wear of the refractories in the upper sections coming into contact with the steel and slag during vacuum treatment. It is essential therefore that the thickness of slag be reduced to a minimum.

A wide range of refractories including aluminosilicate, magnesite–chromite, and fusion-cast zirconia etc., are being tried for the vacuum treatment of refractories in conjunction with the continuous casting process mentioned above.

## REFERENCES

Aristov, G. G. (1969). 'Refractories for Steel Casting' (in Russian), 2nd edn. Metallurgiya, Moscow, pp. 72–92.

Bakker, W. T. and Snyder, G. E. D. (1971). *Bull. Amer. Ceram. Soc.* **50,** No. 3, 235–41.

Brooks, S. H. (1968), *Refractories J.* No. 1, 2–14.

Chesters, J. H. (1963). 'Steelplant Refractories', 2nd edition with Oxygen Addendum. United Steel Company, Sheffield.

Herron, R. H. and Smothers, W. J. (1969). *Bull. Amer. Ceram. Soc.* **48,** No. 5, 544–8.

Humphreys, D. E. (1969). *Refractories J.* No. 5, pp. 128–36.

Ignatov, T. S., Flyagin, V. T. *et al.* (1964). *Ogneupory* No. 7, 33.

Kuznetsov, A. F., Sham, P. I. *et al.* (1970). *Ogneupory* No. 2, 35–9.

Mackenzie, J. (1966). *Refractories J.* No. 4, 134.

Neely, J. E., Boyer, W. H. and Martinek, C. A. (1970). *Bull. Amer. Ceram. Soc.* **49**, No. 8, 710–13.

Nepsha, A. V. (1967). 'Converter Tar-Bonded Dolomite Refractories' (in Russian). Metallurgiya, Moscow, pp. 103–9.

Parnham, H. (1966). *Refractories J.* No. 9, 366–83.

Pitak, N. V. *et al.* (1968). *Ogneupory* No. 7, 28–37.

Sarjant, R. J. (1958). In: *Electric Furnaces*, edited by C. A. Otto. George Newnes, London, pp. 28–63.

Snow, R. B. (1969). *Bull. Amer. Ceram. Soc.* **48**, No. 11, 1042–7.

# Chapter 20

# Use in Non-ferrous Metals

Less than 10 % of all refractories produced in the world are used in furnaces which produce non-ferrous metals: copper, nickel, zinc, tin, lead, aluminium, magnesium and their alloys. However, the employment of refractories for smelting and refining these important metals is obviously very critical, and the development of new furnaces and smelting techniques calls for new advances in refractories technology.

As in the steel industry, the non-ferrous industry, especially as regards copper smelting, has seen increased use of oxygen. This has in turn led to higher furnace temperatures and a need for obtaining improvements in the quality of refractory materials. High-alumina refractories are now replacing fireclay, and chemically bonded and direct-bonded bricks are also finding increasing use. For example, phosphate-bonded castables are finding use in some non-ferrous furnaces, and intensive research is being done to discover more about the complicated reactions that occur between non-ferrous metals and refractory linings and the way linings may be mechanically torn off the furnace housings (*see* Figs. 20.1 and 20.2).

## FURNACE TYPES

In the smelting of many non-ferrous ores the main types of furnace are various forms of the *blast* furnace and the *reverberatory*. Lead is mainly smelted in the blast furnace, while copper is produced generally in the reverberatory, using material in the form of mattes (*see* below). Other furnaces widely used include the *rotary*, which is employed, for example, to recover mercury from its sulphide ore (the *Waelz* furnace is of a rotary design similar to the cement kiln and this is also being used to calcine ballclay and other clays to make chamottes). The name *Waelz* is derived from the German verb wälzen, to roll, a description referring to the motion of the charge in the furnace. *Converters* are metal cylinders lined with refractories and capable of being turned on trunnions for loading and unloading.

FIG. 20.1.   Tensile strength apparatus. *Photo:* Courtesy British Ceramic Research Association.

The particular design of the furnace and the nature of the refractory linings often depends on the location of the raw materials used in the metal refining process. Thus, widely different furnace designs are employed in different countries of the world.

## COPPER SMELTING

The techniques employed in smelting copper can be used as examples for describing the use of refractories for the treatment of other non-ferrous metals. Several types of furnace are used in the production of

FIG. 20.2.   Specimens after testing for tensile strength. *Photo:* Courtesy British Ceramic Research Association.

copper, including the blast furnace, converters and the reverberatory, with the latter having become more prominent because of its greater ability to handle flotation concentrates which constitute the main source of copper at the present time. Copper smelting in a reverberatory furnace requires that the refractories should withstand the chemical corrosion of fluxes, slags, chlorine, steam and sulphur dioxide. Silica brick has been the chief refractory meeting most of these requirements. Today magnesite is used in most roofs (suspension type). The walls of the reverberatory furnace are usually made of chrome–magnesite or silica. Some furnaces operate with the inside hot face made of chrome–magnesite and the cooler outside from silica. The hearth is often made of magnesite while in some furnaces it is rammed with quartz powders or built of silica blocks.

Reverberatory smelting of copper involves the oxidative firing at temperatures of about 1 550°C of a batch consisting of raw or roasted concentrates and fluxes. The batch contains sulphides, oxides, sulphates and carbonates, and so there is ample opportunity for complex reactions between charge and refractory. Many of these reactions have not been studied in detail but researchers are beginning to catch up with the problem. McPherson (1969), for example, studied the changes in the properties and structure of chrome–magnesite refractories following magnesium sulphate formation in reverberatory furnaces and converters. He found that the $MgSO_4$ resulting from the action between periclase and sulphur dioxide and oxygen which had diffused into the zone of the refractory in which the sulphate is stable can cause the brick to crack.

There is evidence for believing that chemically bonded brick resists the formation of magnesium sulphate better than does fired refractory owing to its lower gas permeability.

Aizenberg (1970), concerned with analysing the phases of aluminous and other refractories after service in copper furnaces, points out that no methods have been published for determining copper aluminates in spent refractories. Recent work has resulted in a method for the phase analysis of high-alumina refractory after use in copper smelting furnaces by which it is possible to determine the contents of copper aluminates, mullite, corundum and glass phases.

The products resulting from smelter reactions consist of mattes and slags. Copper mattes are fusions or alloys of copper and iron sulphides and contain magnetite, zinc sulphide, free metals and slag inclusions. According to Dennis (1961) the reaction can be represented as follows:

(1) $2CuFeS_2 \rightarrow Cu_2S + 2FeS + S$

(2) $3Fe_2O_3 + FeS \rightarrow 7FeO + SO_2$

(3) $FeO + SiO_2 \rightarrow FeO.SiO_2$

## Zone formation

A demolished silica roof from a copper reverberatory furnace usually has zones: unchanged, transition, and the hot face or working zone. The transition zone contains a high proportion of glass phase and quantities of wollastonite and fayalite. The working zone consists of fragments of cristobalite, tridymite and quartz together with iron glass, magnetite and other more complex minerals. The phase and chemical compositions of these zones depend on the type of batch being smelted. Zones also form in chrome–magnesite roofs. The hot face reaction, for example, consists of a dense crystallised spinel structure, cemented by silicates such as forsterite and monticellite. Zincite may also be present with cuprites.

Unfired magnesite–chromite is now being used for the roofs of copper smelters. Blast furnaces used for copper smelting are lined with firebricks which should be dense and highly abrasion resistant.

In Russia copper converters which use oxygen-enriched air for the blow, with consequently higher temperatures, are made of magnesia-spinel refractories. Operators have found that chrome–magnesite and ordinary magnesite–chromite brick are unstable in these conditions.

## Service of refractories in copper converters

The reagents acting on copper-converter linings are copper, its oxides and slags. The temperature in the working zone varies in the range 1 200–1 500°C. A zoned structure results from the migration into the refractory of silica, ferric oxide and copper oxides. The chief mineralogical changes taking place during the use of magnesite refractories in copper converters have been studied by several researchers. The most destructive agent is fayalite formed by the oxidation and slagging of iron sulphides. Following the reaction of periclase in the refractory with the slag, fusible compounds of the olivine type are formed. These are enriched with fayalite which wears away the refractory. Copper compounds in the melt speed up the destruction of the refractory. A particularly adverse effect of copper compounds is manifest as the acceleration of the conversion of periclase

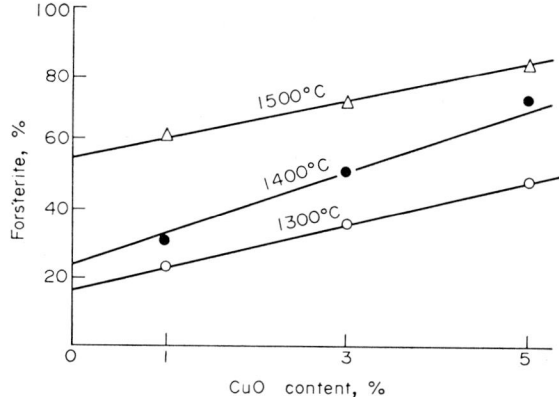

FIG. 20.3.    Effect of copper content on the rate at which forsterite is formed in magnesite linings of copper converters. After Ragozinnikov *et al.* (1967).

into forsterite through various metastable compounds, and then into the low-temperature silicate phase. When chrome refractories are used in copper converters, the copper forms compounds of the type $Cu_2O$. $Cr_2O_3$, which reinforces the wearing of the lining.

Figure 20.3 shows how copper helps to form forsterite in magnesite linings.

When magnesite–chromite linings are employed in copper furnaces it is thought that the copper oxides react chemically with the periclase and olivine bond of the lining to form solid solutions, and the chrome–spinel forms new compounds of the spinel type: $Cu_2O.Cr_2O_3$ and $Cu_2O.Fe_2O_3$ (*see* above).

Rigby and Hamilton (1961) found that a low-temperature fusion migrated into the cooler zones causing cracking and spalling of the basic brick.

**Electric furnaces**
These are also used in the production of copper and various materials have been tried for the linings. Low frequency induction furnaces, for

example, may have bottoms rammed with dry quartz powders containing borax or boric acid as mineralising and consolidating agents. The linings are subjected to fast moving copper at 1 300–1 400°C or above and the conditions are generally so harsh that furnace lives are only 1–3 months.

Recently high-alumina ramming bodies have been used in the USA and elsewhere for induction copper furnaces. These have very low shrinkages and are based on coarse-grained 60–70% alumina chamotte and fine milled alumina–clay materials. The chamotte is made by calcining fine milled alumina and china clay (40:60) at temperatures of 1 620–1 640°C. A more recent development is the incorporation of phosphate bonds. The resulting compositions (containing phosphoric acid) are less porous and gas-permeable and chemically much stronger than bodies not containing the acid. The alumina combines with the acid to form aluminium and silicon phosphates. These materials have lasted up to two years in copper smelting induction furnaces.


## NICKEL SMELTING

Nickel production involves the use of two main furnace types: reverberatory and converters. The refractories supplier has a further interest in blast or shaft furnaces, electric furnaces, rotary kilns (roasters) and electrolytic furnaces.

The nickel extraction process may involve smelting sintered concentrate in a blast furnace, followed by converter treatment of the matte product. Several versions of the process are used, depending on the location of the nickel ore deposits.

Nickel slags often have high magnesia concentrations (up to 15%) and thus high smelting temperatures are required in blast furnaces handling these ores. Firebricks of various classes are used as the lining materials in nickel smelting blast furnaces.

The conditions in nickel smelters are in many ways similar to those in copper smelters and the refractory service is therefore similar. Refractories tried in copper furnaces are in use for nickel processing. Electric furnaces used for cupro-nickel ores, agglomerates and concentrates are built with fireclay, magnesite and magnesite–chromite refractories.

The bessemerisation (conversion) of nickel mattes is done at temperatures of 1 200–1 300°C, using chrome–magnesite refractories. Magnesia–spinel brick has also been used. The chief causes of refractory wear are scaling, slagging and erosion due to the circulating charge.

Zones are formed in the lining, the transition zone containing chromite and periclase, and the hot contact zone consisting of these two minerals and olivine and metallic inclusions. The presence of olivine under these conditions suggests that it is formed by the reaction between fayalite slag and magnesia at temperatures of about 1 300°C (*see* above, under copper smelting).

## ZINC SMELTING

Zinc is smelted in conditions which are extremely severe on the furnace refractories (*see* also *blast furnace refractories* in Chapter 18). Zinc oxide is reduced by carbon and as this is a strongly endothermic reaction as follows:

$$ZnO + C = Zn + CO—57 \text{ kcal}$$

it is necessary to heat the furnace to high temperatures (1 300°C) in a closed cycle to prevent oxidation of the zinc thus produced. The reduction process commences at 960°C and occurs very quickly at 1 100–1 200°C.

The refractories used in zinc smelting retorts must be resistant to corrosive zinc vapours, have a low gas permeability and a high mechanical strength. Usually retort failure is due to the chemical action of the slag and zinc oxide. Upon reaction with chamotte, for instance, zinc oxide forms the spinel $ZnO.Al_2O_3$, which is fusible and corrodes the retort.

The atmosphere inside a zinc retort is reducing and consists of carbon monoxide and zinc vapours.

Silica brick used in zinc smelting (*e.g.* in the roof, arches, or dividing walls) becomes saturated with iron, aluminium and zinc oxides and this rapidly impairs its refractoriness-under-load, especially where the slag contacts the refractory. In the zones near the gas ducts the refractory undergoes severe thermal shock (from 900 to 1 400°C) and with silica structures a large volume change occurs owing to the change in the crystal form (from quartz to metacristobalite). Tridymite may also be formed, causing an additional volume change. The result is that silica refractories do not last long in these places in the furnace. Silicon carbide is used in some furnaces and its high thermal conductivity and spalling resistance give longer lives than silica.

Magnesite refractories are used in the construction of exothermic zinc furnaces which are charged with a mixture of 60% zinc agglomerate, limestone and coke. Other refractories employed are fireclay, magnesia–spinel, and chrome–magnesite. Silica is used for the roof.

## WAELZ PROCESS

Firebrick is usually employed to line the outer layer of zinc distillation rotary furnaces together with an inner layer of chrome–magnesite or other basic refractory. This is the basis of the Waelz process (*see* above) which is also used for producing lead, tin and antimony. A similar setup has been modified to produce chamotte from ballclays. The process used in Britain is basically one of feeding batch into the cold end of a sloping refractory lined tube which is heated at the other end by an oil burner, and discharging the calcined material near the hot end for further processing.

When used to produce zinc, the Waelz process does not yield a product whose quality is high enough for sale as it stands and further treatment is required.

Ringing can be a serious disadvantage with tube furnaces of the Waelz and rotary cement types. This takes the form of a build up of accretions of materials on the walls of the rotating tube. Once they begin to form it is very difficult to break them off the lining and the furnace may have to be halted while the obstructions are removed. It is necessary to control the temperatures, the draught and the chemical composition of the charge and to examine frequently the inside of the furnace if ringing is to be avoided.

In Britain one clay producer has got round the problem of ringing by installing closed circuit television to keep a continuous watch on the furnace interior (*see* Chapter 6). It is likely that the type of lining used in Waelz furnaces has an effect on the formation of ringing.

Chrome–magnesite bricks employed in zinc Waelz furnaces become saturated with iron oxide and silica, the melts migrating into the brick. The absorption of iron oxide may cause the chromite grains to burst and this causes scaling. Other basic bricks, *e.g.* periclase–spinel, are now replacing chrome–magnesite in some countries for Waelz furnace construction.

## TIN PRODUCTION

The reverberatory furnace is the main unit in tin smelting: a process which may involve the production of tin silicates or calcium stannates (depending on whether acid or basic fluxes are used). These compounds form part of the slag which is usually reprocessed to extract the tin. Thus, tin production as it interests the refractories technologist involves: (a) reducing concentrates with coal, (b) recovering tin from slags and (c) refining the pure tin.

The blast furnace and the electric furnaces are also used in tin smelting. In the latter carbon is used for the bottom and walls up to the slag level and silica in the roof. Reverberatory furnaces have hearths built of firebrick as the foundation, on top of which are laid courses of magnesite bricks and magnesite ramming powder. The three materials form a bed about 80 cm thick.

Blast furnaces are rarely used for tin smelting and are quite small. The refractories used are firebrick (for the shaft and throat) and magnesite (bottom and hearth). The atmosphere is reducing and the temperatures may rise to 1 200°C. Iron crusts choke up the lining and cause the furnace to be stopped every 3–5 months.

## LEAD PRODUCTION

Blast furnaces, ore-hearth processes and reverberatory units are all used to produce lead but the blast furnace is probably the most important in terms of total output.

The shaft of the blast furnace is built of firebrick; the hearth is rammed solid with chamotte–clay body and sometimes with magnesite. The lead oxide in the sintered charge is reduced by carbon monoxide which is itself formed by burning coke in air. As in iron smelting, oxygen-enriched blow is used to accelerate the reduction. The temperature in the lead blast furnace varies from 200 at the bell to 1 200°C in the tuyere zone.

When lead is being converted (bessemerised) foams are obtained containing zinc and noble metals. Graphite retorts are housed in a firebrick rocking furnace to distil the foam at 1 200°C, the aim being to separate the zinc by distillation.

Magnesite and firebrick are the refractories used in cuppellation furnaces in which retort bullion is treated to extract an alloy of silver and other noble metals. Oxygen or air is blown through the melt and so very high temperatures are generated. The hearth of the cuppellation furnace is strongly attacked by lead oxides. The refractories are dolomite, aluminous cement concretes, calcium phosphates and firebrick. High-alumina refractories of the mullite type are now being used for such furnaces, especially in the roofs on which lead may be deposited to form highly fusible lead silicates which soon cause the structure to fail in service.

## ALUMINIUM PRODUCTION

In several aluminium producing countries bauxites and other high-alumina minerals are calcined at 550–600°C to produce a dehydrated material which is dissolved in hot caustic solution under pressure. This liquor, after the separation of iron oxide and other impurities, is seeded with hydrated alumina to precipitate some of the alumina which is calcined in rotary kilns at 1 150°C.

The rotary kilns are lined with aluminosilicate refractories such as 42% firebrick and high alumina (70% $Al_2O_3$) mullite, sillimanite, etc. The linings must be highly abrasion resistant because bauxites and other aluminous ores are very harsh on lining wear.

Another process used in aluminium production involves calcining the bauxite with sodium carbonate and lime to produce a sinter, consisting mainly of sodium aluminate. The temperatures of the material are about 1 250°C and of the kiln gases 1 650°C. Slagging and abrasion are the main causes of lining wear.

Aluminium is generally produced by electrolysing fused salts, that is, mixtures of alumina and cryolite. The linings of the bottoms and other parts of these electrolysers are made of several courses of firebrick and a layer of carbon ramming, followed by 3–4 layers of carbon blocks. The walls of the electrolysers are lined with carbon slabs backed with fireclay or china clay insulation. The life of electrolysers (usually up to three years) depends on the quality of the hearth which has to stand the molten aluminium pressure and the chemical attack from these highly corrosive liquids. Magnesite is now being used for constructing walls in electrolysers.

Rotary kilns used for sintering bauxites at temperatures of 1 300–1 400°C in the production of alumina for metal smelting and as raw materials for the refractories industry, have been lined with fireclay brick. The consumption of refractory varies from 2 to 6·3 kg of brick per ton of aluminium, with the linings lasting only 4–6 months.

Bauxite sinter (which may contain sodium aluminate, sodium ferrite, thenardite, dicalcium silicate, $5CaO.3Al_2O_3$, and other complex compounds) contains large quantities of silica. The linings of the rotary kiln may be prematurely destroyed because of the lack of balance between the chemical composition of the material being sintered and the lining. A zoned structure develops, the main constituent of the reaction zone being carbon and nepheline, and small amounts of helenite.

It appears from work done by Zinov'eva et al. (1969) that the most active agent in these conditions is sodium sulphate which forms a lot of glassy foam, attacking the lining almost to the base. These authors recommend basic refractories for lining bauxite sinter rotary furnaces. Tests with chrome–magnesite, for example, have given a life of two years.

## MAGNESIUM PRODUCTION

Magnesium is produced by thermic reduction of the oxide and electrolysis of fused chloride. Production of magnesia from sea water is itself of interest to refractory users since magnesia of this type forms the basis of many furnace linings. Firebrick is used to line the furnaces in which magnesite and magnesium chloride are roasted at about 500–550°C, and also in magnesium electrolysers. Other types of refractories employed in magnesium production include magnesia itself, carbon and also high-alumina products.

# REFERENCES AND BIBLIOGRAPHY

Aizenberg (1970). *Ogneupory* No. 1, 49–52.

Dennis, W. H. (1961). 'Metallurgy of the Nonferrous Metals'. Pitman, London, p. 96.

Gupta, M. M. and Nadachowski, F. (1969). Investigations of the bonds used in SiC refractories for the aluminium industry, *Central Glass and Ceram. Res. Inst. Bull.* **16**, No. 3, 80–3.

Ignatova, T. S., Kudryavtseva, T. N. *et al.* (1968). Mechanism of wear of aluminosilicate refractories by tin and its oxides, *Ogneupory* **33**, No. 9, 23–32.

McPherson, R. (1969). *Bull. Amer. Ceram. Soc.* **48**, No. 8, 791–3.

Ragozinnikov, V. A., Shchetnikova, I. L. and Vorob'eva, K. V. (1967). Physicochemical processes in the destruction of refractories in copper metallurgy, in: 'Chemistry of High-Temperature Materials'. Nauka, Leningrad, pp. 213–6.

Rhee, S. K. (1970a). Wetting of ceramics by liquid aluminium, *J. Amer. Ceram. Soc.* **53**, No. 7, 386–9.

Rhee, S. K. (1970b). Wetting of AlN and TiC by liquid Ag and liquid Cu, *J. Amer. Ceram. Soc.* **53**, No. 12, 639–41.

Rigby, G. and Hamilton, B. (1961). *J. Amer. Ceram. Soc.* **44**, No. 5.

Rutman, M. M., Cherepok, B. V. *et al.* (1968). Resistance of aluminosilicate refractories to aluminium alloys, *Ogneupory* **33**, No. 11, 37–40.

Tseitlin, L. A. and Merkulova, E. V. (1967). High-alumina ramming bodies for copper induction furnaces, *Ogneupory* No. 8, 34.

Uglikova, N. S. and Perepelitsyn, V. A. (1968). Testing bonds of refractory concretes in molten zinc wastes, *Ogneupory* No. 1, 26.

Zinov'eva, N. V., Ragozinnikov, V. A. *et al.* (1969). *Ogneupory* No. 4, 35–7.

## Chapter 21

# Refractories in Cement Furnaces

Today the unit most commonly used to make cement clinker is the rotary cement kiln. The production and properties of refractory materials for the Portland cement industry are therefore chiefly concerned with meeting the conditions prevailing inside such furnaces which are tubular in shape, about 6 feet in diameter and up to 500 feet in length.

Raw materials in the form of clay-like minerals and lime minerals, either blended as a slurry or dry, are fed in at one end of the rotary kiln.

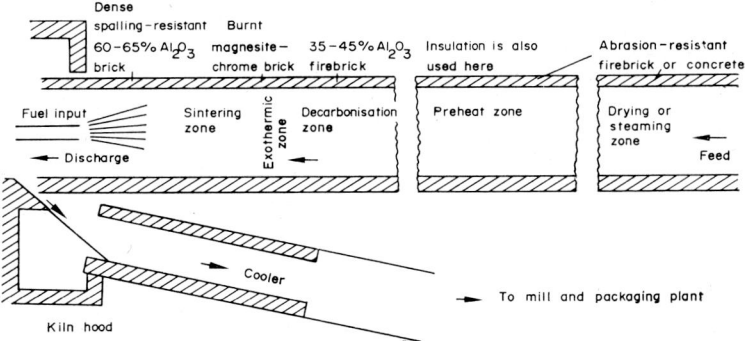

Fig. 21.1.   Zones and refractory linings in a rotary cement kiln.

Fuel such as powdered coal, oil or gas is burned at the other end of the kiln so that the hot gases move in the opposite direction to the raw materials. The kiln itself is inclined slightly to the horizontal and during firing can be rotated, so that the materials inside gradually move along the tube. In other words, raw materials enter at one end and burnt clinker leaves at the other. The action of the kiln is to blend and burn the raw materials, processes during which complex physicochemical reactions occur, converting the clays and limestones into polymineral grains of Portland cement clinker about 10–20 mm in size.

Ordinary Portland cement is made from about 75–80% lime minerals and 20–25% clay constituents. The most critical and corrosive reactions take place in the sintering zone and here the refractory life depends on the formation of a protective skin. This skin is formed by the reaction between the lining refractories and the raw materials fed into the kiln. In order to foster the development of a stable skin, materials other than limestones and clays are used in the feed. They include roasted pyrites, flue dust from blast furnaces and tripoli: all with high iron concentrations.

Portland cement is a hydraulic setting material with a chemical composition in the following limits: 64–67% $CaO$, 1–6% $Fe_2O_3$, 20–25% $SiO_2$, 0·5–3·5% $MgO$, 4–8% $Al_2O_3$, and 0·3–1·0% $SO_3$. In magnesia cements the $MgO$ content may exceed 8%. Commercial clinkers also contain 0·1–1% alkalis, $Na_2O$ and $K_2O$.

Special duty cements which may have radically different chemical compositions from that given above are also made, *e.g.* high-alumina cements for the preparation of refractory concretes and castables and barium-aluminate cements for protection against nuclear radiation.

The minerals found in calcined cement clinker include: 35–69% alite ($3CaO.SiO_2$); 9–43% belite ($2CaO.SiO_2$); 2–18% tricalcium aluminate ($3CaO.Al_2O_3$) and 1–18% of $4CaO.Al_2O_3.Fe_2O_3$.

A rotary cement kiln may be roughly divided into five zones (Fig. 21.1) in which specific cement-forming reactions occur. These zones, the conditions prevailing in them and the types of refractories employed to make the linings will now be discussed.

## STEAMING AND PREHEAT ZONE

The water contained in the slurry fed to the kiln is evaporated and the temperature gradually rises to 400°C. The slurry becomes crumbly and sluggish. Chains are suspended from the inside of the lining to help break up these lumps. Approaching the preheat zone where the temperature is 450–750°C, the material loses its organic matter (at 450–500°C); the clays are dehydrated (that is, they lose their water of crystallisation) and the kaolinite molecule decomposes. The fine pellets are changed into a powdery mass.

Refractories for this zone need to be highly abrasion resistant in order to resist the flow of materials and the impact of the chains. The PCE value is not important. Sometimes the zone is not even lined with a refractory lining but if it is, firebrick is suitable. Since the feed is highly viscous, subsequently becomes powdery, and tends to cling to the lining, any design provision that will help push the slurry through the kiln is to be welcomed. Thus, castables in the form of heat-resistant concretes which eliminate the need for brickwork joints and encourage easier flow are being used for lining this zone.

Clinker-cement concrete used to line the entrance zone of the kiln and the adjacent chain-curtain section is prepared when the rotary kiln is being lined, using clinker and cement that have been previously prepared in the normal way. Well burnt, stabilised clinker not containing any free lime, together with aggregate grains of 3–8 mm and a moderately high grade of Portland cement should be used to make the lining.

Heat losses in the preheat zone can be reduced by using insulating (fireclay or high-alumina lightweight) materials having a bulk density of about $1 \cdot 3$ g/cm$^3$ and a crushing strength of at least 100 kg/cm$^2$.

## DECARBONISATION ZONE

The temperatures here may reach 1 200°C. Carbonates are decomposed and sulphur oxides and any residual organic material are removed from the batch materials. Incipient solid-state reactions between the finely dispersed lime, silica and sesquioxides ($Al_2O_3$ and $Fe_2O_3$) are in evidence. The powdery mass changes into a coarsely granular material. Suitable refractories for this zone are firebrick containing 35–37% alumina for use in the less critical parts of the zone, followed by 44%-$Al_2O_3$ firebrick for the higher temperature end of this section. The thermal-shock resistance and abrasion resistance are important properties of materials used in this part of the furnace.

Lightweight siliceous insulating blocks are also used with varying degrees of success. Low-alkali slurries give better results than high-alkali materials because alkalis develop fusible compounds in the lining which lead to rapid melting of the lining materials.

## EXOTHERMIC ZONE

Vigorous solid state reactions occur here at temperatures around 1 400°C. More heat is generated than is consumed by the chemical reactions, and so the temperature rises quickly (over a distance of about 3–4 metres in the rotary kiln). The main reactions are as follows:

At 700–1 000°C alumina is bonded into $CaO.Al_2O_3$, and small amounts of $2CaO.SiO_2$ are formed.

At 1 000–1 200°C the $CaO.Al_2O_3$ is saturated by lime to form $5CaO.3Al_2O_3$, and then $3CaO.Al_2O_3$.

The ferric oxide reacts with calcium oxide to form $2CaO.Fe_2O_3$ and $4CaO.Al_2O_3.Fe_2O_3$. The formation of the compound $2CaO.SiO_2$ is completed and the excess calcium oxide remains as free lime.

A serious problem in this zone is frequently met with in the form of 'ringing' or 'ring formation', causing production troubles and loss of efficiency. The precise causes of ringing, which also occurs in rotary kilns burning clay into chamotte (grog), are not clear. The use of basic bricks in place of high-alumina products has been suggested as one possible cure. Anything that tends to disturb the regular operation of the kiln, such as a change in the properties of the feed, draught or fuel, may suddenly produce ringing problems. The rings, which block the cross section of the kiln and disturb the aerodynamics, are formed from ash or from the materials being fired.

According to some kiln operators, the composition of the cement feed materials does not greatly affect lump formation (leading to ringing). Frequent examination of cement kilns shows that at the moment of formation of the lumps there is no skin at the start of the sintering zone. When the lumps are removed, a ring up to one metre thick appears (4–5 times thicker than in the centre of the sintering zone). The ring-shaped skin is from 1–3 metres long and the internal layers are gradually cooled. Cracks eventually appear in them and the rings break down. This process is periodically repeated. Lumps of material weighing up to two tons and measuring 1–1·5 m across have been observed. The effect seems to be one of 'snowballing'. It would appear that the ringing fault is the cause, and not the consequence, of snowballing in cement kilns. The rate of growth of the rings depends on the size of the kiln. Ash depositions are another critical factor.

## SINTER ZONE

Known also as the burning zone, here the material reaches a temperature of 1 400–1 450°C, falling at the end of the zone to about 1 300°C. The materials are finally burnt into Portland cement clinker in this zone. The essential features of the zone are those of fusion and recrystallisation. At temperatures of about 1 300°C the material is partly fused, liquid phases appearing in the form of $4CaO.Al_2O_3.Fe_2O_3$, $3CaO.Al_2O_3$, MgO and CaO. As the temperature rises this liquid dissolves the free lime and much of the dicalcium silicate, the latter being saturated with lime to form alite. The mass is densified to form strong clinker granules.

The physicochemical reactions taking place in this zone are critical for the formation of the protective skin on the sinter zone lining. Here the lining is subject to the chemical reactions going on in the clinker and to the combustion products from the fuel. Because of its high content of lime and the presence of a liquid phase containing chemically active dissolved CaO, the sintering clinker is very corrosive towards the refractories of the lining, regardless of what type of material is being used.

Another adverse factor in the sintering zone is the temperature gradient through the lining: from 1 450 to 300°C. As the kiln rotates its lining first receives the action of hot kiln gases at a temperature of 1 600°C, and then is coated with a layer of clinker at 1 400–1 450°C. The conditions are very harsh.

Refractory materials for the sinter zone must be chemically basic because of the high calcium environment. In addition to the obvious properties of high fusing point and refractoriness-under-load and also a high flux resistance, the sinter zone lining must also be volume stable, have a high thermal-shock (spalling) resistance, and must be able to foster the formation of clinker skin to protect itself from further attack. Protective skin formation in rotary cement kilns is most important.

Yount and Powers (1969) found that for aluminosilicate refractories and grey cement the extent of skin formation on a model lining depends on the silica–alumina concentration. Only thin skins are formed on firebrick because of excessive reaction: whereas on very high-alumina (90%) material no skin is formed because of lack of reaction. Unstable skins can be obtained with 70–80% alumina refractories. The thickest skins can be produced when the original materials contain substantial amounts of dicalcium silicate or calcium ferrite. These model-scale experiments included the use of special direct-bonded bricks which offered promise as cement kiln lining materials. Forsterite and chrome refractories spalled too quickly.

Expressed simply, the process of skin formation starts when nodules of clinker liquefy. The liquid cooled by contact with the cooler lining is solidified and bonds the solid grains of raw material feed with the lining, thus forming a skin.

According to some British producers (e.g. Jones, 1969) a satisfactory refractory for good skin formation and other cement kiln conditions is burnt magnesite–chrome bonded radially with steel sheets, which when oxidised to $Fe_3O_4$ reacts with the magnesia in the lining and yields a monolithic magnesio-ferrite bond.

Other refractories used in various countries for sinter zones include: high-alumina, magnesite–dolomite, chrome–dolomite, forsterite, etc. In Austria, Canada and Finland, insulating linings behind the sinter zone have been made as a bed of chamotte, and also from double-layer brick. Insulation is also used in the sinter zones of Russian cement kilns. The subject of insulation in sinter zones is one on which many conflicting views have been expressed.

For some small rotary cement kilns producing white clinker the sinter zones may be lined with natural blocks of talc–magnesite stone which lasts about 60 days. Dolomite linings which are claimed to be as good as fired magnesite–chromite may also be important for producing very white clinker because of the lack of chromium oxide contamination, and lower costs.

Factors other than the quality and composition of refractories obviously affect kiln lining life. Whether the linings are layed with mortar or with plates is one such factor. Large kilns should be lined with mortar. Magnesia–iron mortars, containing sodium silicate are being tried, as well as conventional refractory aluminate compositions.

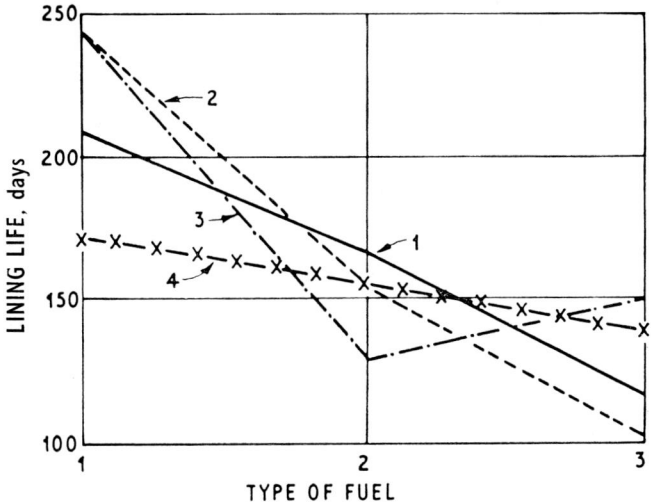

FIG. 21.2.    Lining life of rotary cement kilns using various refractories as a function of the type of fuel used: (1) gas; (2) fuel oil; (3) coal. Materials: (1) Chrome–magnesite lining; (2) magnesite–chromite; (3) unfired magnesite–chromite; (4) periclase–spinel.

The type of fuel used will also affect lining performance. Figure 21.2 shows lining lives of various Russian-made refractories versus gas, oil and coal fuels.

Yet another critical factor assuming increasing importance with the construction of large diameter kilns is the size of the rotary kiln. Figure 21.3 shows the connection between lining life and kiln diameter in sinter zones for two of the commonest refractories used: chrome–magnesite and magnesite–chrome. Shaw (1970) has reviewed the performance of rotary cement kiln linings in the Soviet Union.

## COOLING AND DISCHARGE ZONE

High-spalling resistance is the main property of refractories used in the discharge zone, since the lining undergoes sudden fluctuations in temperature: from the clinker at 1 400–1 500°C to secondary air contact at

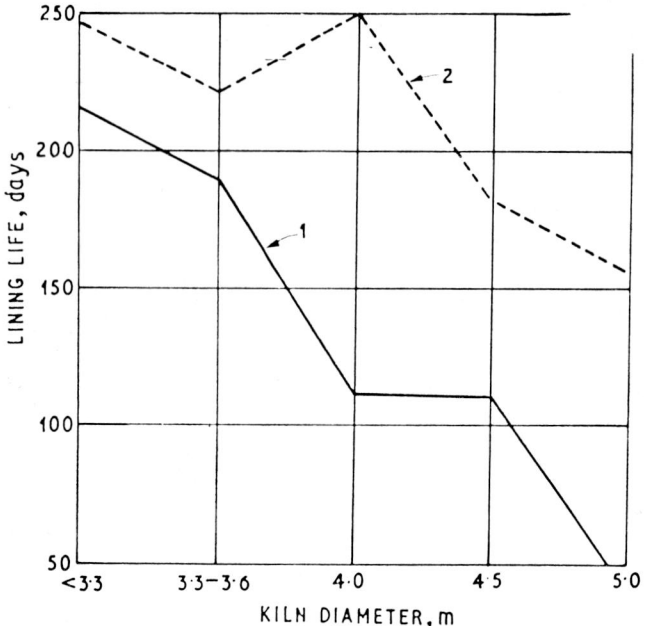

Fig. 21.3. Relationship between the lining life and cement kiln diameter in the sinter zone. (1) Chrome–magnesite refractory; (2) clad magnesite–chromite refractory.

200–400°C. A dense 60–65% alumina brick with a very high abrasion resistance is often used in this zone. In the cooler, the temperature of the clinker is reduced to 1 000°C, part of the liquid phase crystallises, and the rest solidifies (forming a vitreous mass).

The various lining zones overlap, and no strict divisions can be laid down to cover all sizes of furnace. However, as a general rule the following figures give some guidance:

| | |
|---|---|
| Steaming and preheat zones | 14–25 kiln diameters (or half the total length) |
| Carbonisation and exothermic zones | 7–12 kiln diameters (one quarter of the length) |
| Sintering zone | 4·5–6 diameters (15% of the length) |
| Cooling zone | 1–2 diameters |

Recently the trend has been away from old established wet methods of slurry feeding towards the dry or semi-dry methods with consequent fuel and time savings. The length of the steaming zone needs to be much less with the dry method and the above ratios are consequently greatly altered.

## REFERENCES AND BIBLIOGRAPHY

Jones, R. (1969). *Refractories J.* No. 5, 138–43.

Nachtwey, W. and Prost, L. (1969). Development of improved refractory bricks to improve rotary cement kiln linings, *Ind. Ceramique* No. 621, 597–601.

Shaw, K. (1970). Performance of rotary kiln linings in the Soviet Union, *Cement Lime and Gravel* No. 8, 201–4.

Trojer, F. (1968). Deterioration of cement rotary kiln linings by alkali sulphides and sulphates, *Amer. Ceram. Soc. Bull.* **47,** No. 7, 630–6.

Yount, J. G. and Powers, W. H. (1969). Simulated service tests of cement kiln brick, *Amer. Ceram. Soc. Bull.* **48,** No. 7, 716.

## Chapter 22

## Glass Melting Refractories

Glass quality depends very much on the purity and general performance of the refractories used in making glass furnaces. Furnace operating speeds are also closely related to refractory quality since high temperatures mean faster working and better glass, and of course high temperatures depend very much on the quality of refractories.

The refractories employed in glass melting furnaces must resist molten glass and vapours which are highly corrosive. Since the temperatures inside the furnaces may change rapidly, the bricks must also be thermally shock resistant.

The following refractories are employed: fireclay, kaolin, high-alumina (sillimanite and mullite), zircon–mullite, zircon–alumina; magnesite, chrome–magnesite, forsterite. Silica refractories and semi-silica as well as a range of chamotte blended with kaolin and also refractories based on pyrophyllite and zircon are employed. In glass lehrs silicon carbide muffles operating at temperatures up to 600°C for removing stresses by annealing are also employed. Silicon carbide also finds use for recuperators in the furnaces.

TABLE 22.1

*Chemical Composition of Glass Making Refractories*

| Material | Content, % | | | | | | | |
|---|---|---|---|---|---|---|---|---|
| | $SiO_2$ | $TiO_2$ | $Al_2O_3$ | $Fe_2O_3$ | CaO | MgO | Alkalis | $ZrO_2$ |
| China clay blocks | 53·4 | 0·8 | 43·0 | 0·7 | 0·5 | 0·2 | 1·5 | — |
| Siliceous brick | 71·2 | 0·5 | 25·8 | 0·4 | 0·4 | 0·5 | 1·2 | — |
| Fireclay brick | 56·9 | 1·4 | 36·5 | 0·8 | 0·8 | 0·5 | 3·1 | — |
| Sillimanite | 27·1 | 1·1 | 60·0 | 0·6 | 0·4 | 0·1 | 0·7 | — |
| Fused mullite | 23·2 | 2·6 | 70·2 | 1·3 | 1·2 | 0·6 | 0·9 | — |
| Zircon–alumina | 14·4 | 0·4 | 62·9 | 0·5 | 0·5 | 0·3 | 0·8 | 20·2 |
| Corhart ZAC | 12·8 | 0·7 | 50·7 | 0·5 | 0·4 | 0·2 | 1·2 | 33·5 |
| Silica brick | 95·0 | 1·1 | 0·5 | 0·3 | 2·3 | 0·1 | 0·4 | — |

## SERVICE OF REFRACTORIES

The stresses in, and chemical attack on, refractory linings in glass furnaces commence as soon as the warming up begins. Since glass tanks (in which the glass is melted) are made of very dense refractories the linings may undergo cracking and spalling damage if the furnaces are heated too quickly from cold. This damage may continue to threaten after the furnace has been in operation for some time, especially in burner ports and regenerators and during the flame change-over (*see* below).

FIG. 22.1.   Suspended silica feeder end wall in a glass furnace—an example of sophisticated design technique. The support is transferred from refractory to the more suitable structural material such as steel. *Photo:* Courtesy M. H. Detrick.

Glass tank refractories are worn out chiefly by chemical erosion due to molten glass and fused salts. The tank may contain salts such as sodium carbonate and sulphate and also lead compounds in the melting of lead crystals. Such materials are highly reactive with most types of refractory. In the flame regions of the furnace batch dust may drop onto linings together with volatilised materials from the melting glass, causing further corrosion of the refractories.

Compositions containing boric oxide, phosphorus, fluorides, barium and lead, as well as highly alkaline glasses, are especially corrosive towards glass tank linings.

Glass is melted by a continuous process (*see* below) and the velocity of glass currents over the various furnace parts may be quite high. This erosive action is another cause of refractory failure.

Recently the glass industry has been concentrating on boosting melting temperatures and is now using ancilliary techniques, such as bubbling air

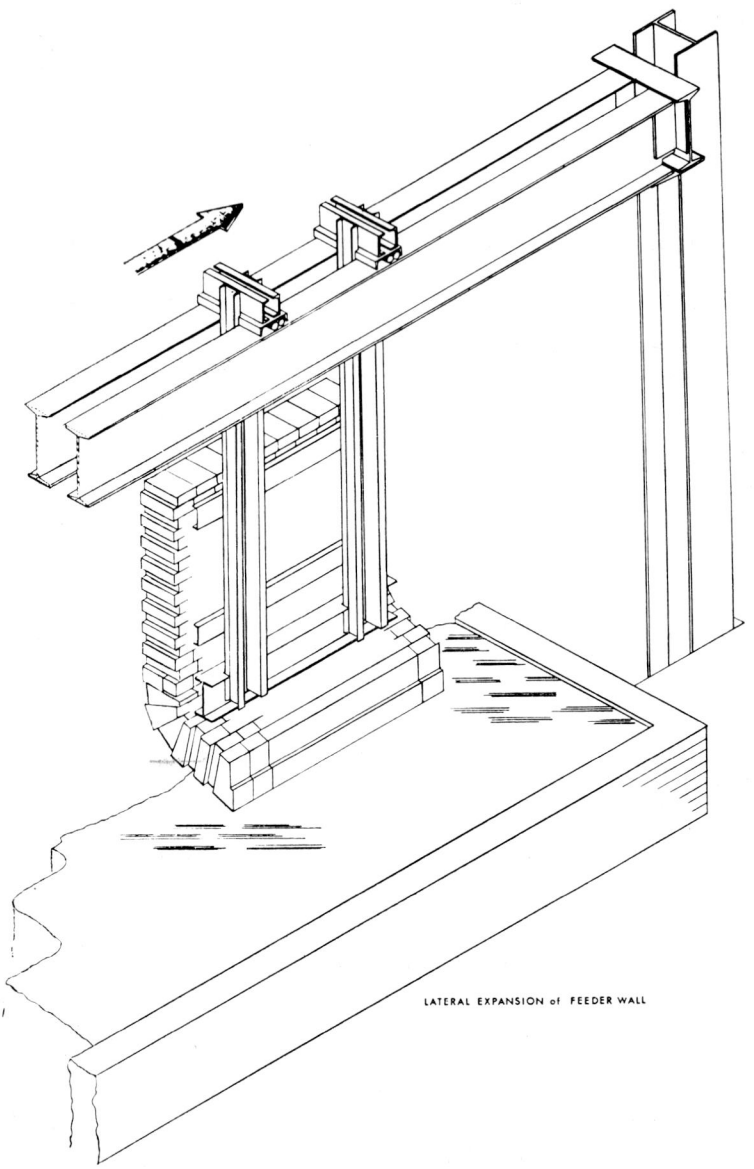

LATERAL EXPANSION of FEEDER WALL

FIG. 22.2.    By providing free lateral movement this feeder wall design compensates for the growth of silica brick during heat-up. The monolithic brick structure has no expansion joints but expands and contracts on roller frames. Courtesy M. H. Detrick.

through the molten glass, all of which are tending to make the life of refractories more severe. It is estimated that an increase in the glass melting temperature of about 50°C halves the life of most tank linings. In fireclay pot furnaces the temperature rise needs to be much less in order to halve the life of the pots.

The greatest degree of corrosion and erosion is noted where the glass surface comes into contact with the refractory lining. Tank blocks are usually made as large as possible to reduce the number of joints since these are another source of weakness.

Dry-laid and mortar-bonded structures are used in all parts of glass furnaces. The mortars have to be specially selected to match the chemical composition of the refractories.

## TYPES OF GLASS FURNACE

Glass is made by feeding a blend of sand, sodium carbonate and cullet (broken glass) and other materials into a tank (melting chamber) where it is fused by heating with oil, gas or electricity, or a combination of gas and electricity. Several versions of this basic process are used. The continuous recuperative furnace is the most common unit in the glass industry. The melting of small amounts of glass, say in experiments or for special purposes (especially for optical glass) is done in refractory pots which are housed in a furnace resembling a beehive around which the pots are located and in the middle of which burns the flame. The molten glass is then poured from the pots into moulds or shaped in other ways. Manual processing is common with this type of furnace.

When the glass materials have been melted the glass flows across the melting chamber into refining sections where it is conditioned ready for forming before passing along forehearths or channels. Sometimes the channels are omitted.

At present numerous experiments are being conducted into glass furnace design, especially in relation to new materials, chiefly fusion-cast and basic refractories which are needed to withstand higher operating temperatures.

### Tank furnaces

Continuous tank furnaces are used for melting most glasses including bottle (containers), sheet (window and plate) and tableware glasses. They are classified in accordance with the direction of the hot gases over the melting glass and the method of dividing the various zones (melting, cooling and working, etc.). Most large regenerative continuous tanks have cross flames and the smaller furnaces have horseshoe shaped flames. The purpose of dividing up the gas regions in the furnace is to achieve better control over heat exchange. Latticed and solid baffles and screens are used for the same purpose.

Cross-fired glass furnaces using natural gas or oil fuels employ the reverberatory principle, that is, the flame licks across the bath of molten glass in one direction and is then reversed, the change being made every 20–30 min.

Booster-electric heating is often used in many types of furnace. It involves the use of electrodes which are fixed through the side walls of the melter below the glass level. Current is fed to the electrodes and the resulting resistance produces extra heat which helps the flames to melt the glass.

The use of electricity in glass melting would undoubtedly become more popular if costs were reduced. However, electrode design and materials are other obstacles to further progress in this form of melting. At present commercial production or trials are proceeding with alkaline, borosilicate and non-alkaline glasses and fluoride and titanium glasses on a very limited scale. Lead glasses for making tableware are a distinct possibility for economic electric melting, but until recently the development was being held up by the problems of lead–molybdenum reactions between the glass and the electrode materials (molybdenum disilicide is commonly used for heating elements and electrodes in the glass, ceramics and metals industries). Tin oxide is also used for electrode making but here again difficulties are encountered with the reliability of the electrical contact. Oxide–tin compositions are being tried to improve these contacts.

Bubblers are now being increasingly used. Holes are made in the melter bottom and air is bubbled through the glass causing homogenisation of the composition and evening out the temperature. This process can adversely affect refractory life.

Horse-shoe flame furnaces are those in which the flame is emitted from oil-fired burners located at one end of the furnace. The gas current is made to swing across the glass bath in a horse-shoe pattern. Recuperators and regenerators need to be used with this type of furnace.

Unit melters are narrow furnaces with burners located on each side above the melt level. The flames from the burners, heated with oil or natural gas, meet in the middle of the batch and form complex patterns of hot-gas currents. The manner in which the currents transmit their heat to the surface of the molten glass, and the resulting patterns of glass current is extremely complicated and is not fully understood. Furnace designers are undecided about the precise advantages of these types of furnace.

## FURNACE PARTS AND SUITABLE REFRACTORIES

The various elements in glass furnace design are subjected to greatly differing conditions of service and so different refractories are used.

In the melting chamber the crown or roof has for many years been built of silica because not only is this a good refractory but when it fuses and drops into the glass it is dissolved without causing discolouration or 'seeds'. For the same reasons it is not practicable to use chrome-containing refractories which, although highly heat-resistant, might cause green staining and seeds. It is possible, however, to employ chrome–magnesite in some parts (*e.g.* the checkers) if green bottle glass is being melted.

FIG. 22.3. Refractory wear at the doghouse (charging end) of glass furnaces is often severe because of the abrasive action of batch. Here raw materials are being fed into a float glass furnace. *Photo:* Courtesy Pilkington Brothers Limited.

In addition to the high temperatures (1 450–1 625°C) glass furnace roofs have to withstand chemical attack from alkalis and aluminosilicates from the batch. Many new furnaces are, therefore, being made with crowns built of high-alumina materials such as mullite. Basic refractories are also being tried chiefly as high-purity magnesite (sometimes direct-bonded). The chemical purity of crown refractories is very important since small amounts of refractory inclusions in the glass can cause high losses in the form of seeds, cords and staining. These faults often do not appear until the glass is being drawn into ribbons or in other shaping processes.

TABLE 22.2

*Relative Corrosion Rates of Refractories by Molten Glasses at 1 480°C*[a]

| Refractory | Relative corrosion rate | |
| --- | --- | --- |
| | Ordinary | Composition containing 2% Na$_2$O and 19% CaO and MgO |
| Zircon–alumina | 0·2–3 | 4 |
| Zircon–mullite | 0·5 | 6–8 |
| Sillimanite and thermite–alumina | 0·6–0·7 | 6–8 |
| China clay | 1·8 | — |
| Firebrick | 2–3 | 4–5 |
| Fused quartz | 5–6 | 0·6 |

[a] After N. V. Solomin, 'Refractories for Glass Furnaces', Stroiizdat, Moscow, 1961.

Furnace walls, including those below the main roof and at the back of the melter forming the superstructure of the furnace, are being increasingly made of fused refractories. Zirconia–alumina and mullite types are popular. Grain-bonded sillimanite and mullite are also in use. Fused zircon (zirconium silicate) is used on its own for some furnace parts which undergo excessive abrasion and high temperatures.

The refining section and the forehearths and feeders are built of high-duty silica, fusion-cast mullite, and more recently of zirconia–alumina refractories. Ports and regenerator chambers have to withstand severe thermal shock as well as high temperatures, and possibly corrosion from fuel ash. Basic refractories are now being used for regenerators.

## REFRACTORY WETTING BY MOLTEN GLASS

One of the chief sources of the destruction of glass furnace refractories is obviously the chemical reaction between glass and lining. However, the

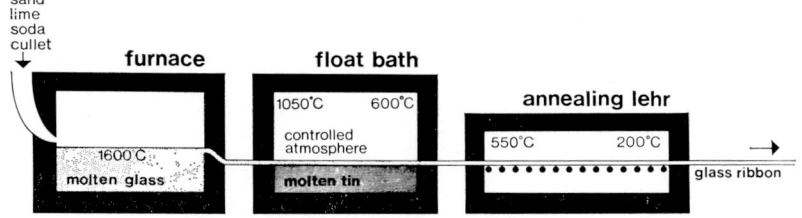

FIG. 22.4. The Pilkington float glass process. Refractories have to be carefully selected not only for melting the glass at 1 600°C, but also for containing the molten tin in a controlled atmosphere. Courtesy Pilkington Brothers Limited.

reaction is not easily explained and as it is difficult to make direct observations (most data is obtained by analysing demolished furnaces), controversy exists even about the precise manner in which glass wets the refractory and about the relationship between wetting action and the physical and chemical properties of the material.

FIG. 22.5. The quality of refractories used in glass furnaces affects the final product. Seeds, inclusion and cords may result from poor refractories or excessive corrosion of furnace blocks. Here a continuous ribbon of top-grade float glass leaves the bath of molten tin and is cooled in controlled stages. *Photo:* Courtesy Pilkington Brothers Limited.

One thing seems fairly certain: the greatest attack occurs at the molten-glass level. The system here involves at least three media: molten glass, refractory, and furnace atmosphere. This partly explains why the degree of interaction is so much more intense than under the glass level where only two phases are relevant: glass and refractory. It may also explain why bubbling air into glass to homogenise it increases lining wear.

Several workers have studied the problem of glass refractory wetting, and Prokof'eva (1971) has recently described studies on how window, lead, and borosilicate glasses wet various refractories. The refractories selected were Russian, American, French and others. They included corundum (96·1% $Al_2O_3$ + $TiO_2$); 99·5% quartz; high-alumina; fire-bricks of various alumina and grog concentrations; kaolin brick, dense fireclay; a range of new experimental products, and French produced ZAC–1681 (zircon) refractories.

The manner in which the glasses wetted the refractories was checked by measuring the wetting angle by the sessile drop method, using the hot microscope. It was concluded that the wetting of refractories is frequently accompanied by absorption, solution or reaction. These phenomena occur with very little distinction within one group of refractories, but may differ markedly in the case of refractories of different types, and, therefore, tend to introduce various distortions to the wetting process, changing the form and size of the droplet. This disturbs any relationship there might be between wettability and destructibility.

The wetting angle depends not only on the chemical compositions of the refractory and glass but also on the porosity and pore sizes of the former.

## REFERENCE AND BIBLIOGRAPHY

Boggum, P. P. (1969). Tank superstructure—basic, *Proc. Brit. Ceram. Soc.* No. 14, 79–91.

Bonetti, G., Toninato, T. *et al.* (1969). Resistance of refractories to corrosion by lead glasses, *Proc. Brit. Ceram. Soc.* No. 14, 29–40.

Busby, T. S. and Carter, M. (1968). Effect of temperature cyling on oxidising and reducing atmosphere on the structure of basic bricks, *Glass Technology* No. 6, 154–63.

Cevales, G. (1970). Corrosion testing of refractories in model tanks, *Glastech. Berichte* **43**, No. 12, 501–3.

Cholerton, J. F. and Charlton, W. (1969). Selection of refractory materials for glass tank regenerators, *Proc. Brit. Ceram. Soc.* No. 14, 59–78.

Henthorn, R. S. and Jackson, B. (1969). Refractories for superstructure and ancillary parts of glassmaking furnaces, *Proc. Brit. Ceram. Soc.* No. 14, 14–57.

Lambert, H. (1968). Insulation of glass-melting furnaces, *Sprechsaal* **101**, No. 22, 1018–28.

Mel'nik, M. P., Firer, M. Ya. *et al.* (1969). Refractories for glass furnace regenerator checkers, *Steklo i Keramika* **26**, No. 4, 11–14.

Prokof'eva, E. A. (1971). *Ogneupory* No. 6, 48–53.

Robijin, P. (1969). Corrosion of refractories by arsenic in glass furnace checkers, *Verres et Refractaires* **23**, No. 2, 201–5.

Schmid, O. (1967). Use of basic brick in the glass industry, *Tonindustrie Zeit. und Keram. Rund.* **91**, 299–300.

*Chapter 23*

# Ceramic Kilns

The ceramics industry itself is quite a large consumer of its own products. Heat-resistant materials are needed for building tunnel and periodic kilns and driers which are used for firing bricks, pipes, electric insulators, tableware, tiles, acid-proof products, vitreous enamels and the refractories themselves. Many of these materials have to withstand very stressful conditions including high temperatures, slag attack from vapours and sulphur deposited on the crowns and walls, fuel ash corrosion, etc. Although it is true that in many countries clean air legislation and the general use of purer forms of fuel such as natural gas, electricity and sulphur-free oils have lessened the severity of slagging attack in ceramic furnaces, the recent drive towards much higher operating temperatures and faster firing cycles has meant that kiln structures have had to be improved in other directions, *e.g.* higher refractoriness, better spalling resistance, and better heat insulating capacity.

## DESIGN OF KILNS

The need for understanding the properties of the ware being fired on the part of the kiln designer is fairly obvious. In order to build a kiln which will convert raw materials into ceramic materials it is essential to understand the complex physicochemical reactions which occur during the heat processing. Some knowledge of the final quality, *e.g.* the brilliance of a glaze or the acid resistance of a lithograph, is essential to the ceramic kiln designer. Usually a prospective firing curve is developed for the particular product on which the kiln design will be based, *e.g.* the length of the various zones, the maximum temperatures to be employed, the soaking times and the cooling times. The kiln atmosphere is also important since the industry employs oxidising, reducing, and neutral atmospheres depending on the type of product being handled. Other factors to be taken into account include the type of fuel to be used and the design and location of the burners. These factors will determine the types of refractories to be employed for constructing the kiln.

Dale (1964) has analysed in detail the decisive factors influencing the choice of kiln design for firing pottery. Many of the principles involved apply also to the firing of other types of ceramics.

The exact choice of refractories for constructing ceramic kilns will depend primarily on the maximum temperature to which the ware is to be fired. For example, the tunnel kiln firing of sillimanite refractories will need much better refractories for the linings of the high-temperature zones than are required in conventional kilns firing common earthen tableware which is fired at 1 050–1 100°C. However, if fast cycles are used with rapidly maturing pottery bodies, the maximum temperature for the earthenware needs to be raised and thus kiln refractories must be of enhanced quality in order to withstand the more demanding conditions.

Taking the ceramics industry as a whole, which uses firing temperatures ranging from about 700°C for the decoration firing through 1 450°C for porcelain and bone china firing, up to about 2 000°C for firing special refractories (usually in small kilns and laboratory units), it is apparent that the whole gamut of heat-resistant materials will be of interest to kiln designers. However, most large kilns are built of firebrick and other aluminosilicate materials, silica brick, magnesite, chrome–magnesite, silicon carbide and various insulating materials.

In recent years there has been a trend toward the use of fusion-cast (including zirconia) refractories, especially in the high temperature zones of tunnel kilns burning structural clay products and refractories.

### Structural clay products firing
Various designs of kiln are used for building bricks, pipes, tiles and other structural clay products. Since suitable clays for making building bricks are to be found in every country of the world, there are dozens of different types of kiln design, determined by the nature of the raw materials and fuels available, the quality of product design and other economic factors. Periodic ovens and tunnel kilns as well as new types of shuttle kilns are in use.

Rowden (1964) has described the kilns used for firing bricks under the following classifications:

1. Clamps.
2. Intermediate kilns (periodic).
3. Semi-continuous kilns.
4. Continuous kilns.

The conditions encountered in each type of kiln and the types of fuels employed are also discussed by this author.

The building-ceramics industry is also interested in the production of chamottes, grogs and expanded clay aggregate. Periodic and continuous furnaces of various designs with varying refractories requirements are used

FIG. 23.1.   The firing of fine tableware requires various types of refractory kiln furniture units. *Photo:* Courtesy Rosenthal-Porzellan AG.

FIG. 23.2.   View of the inside of the Trent kiln for 'one-high' firing of sanitaryware. *Photo:* Courtesy West Midlands Gas Board.

for calcining raw materials for use in making bricks, concrete aggregates, pipes and blocks. For example, shaft and rotary furnaces are used for firing clay into grog (also, more recently, fluidised bed furnaces).

In the tile industry clay powders for pressing are now being prepared by spraying in high-temperature spray driers. High-alumina linings are needed for these driers.

### Pottery firing

Recent trends in the firing of earthenware, faience, porcelain and bone china include faster firing cycles, the use of small cross section kilns, 'one-high' firing and higher firing temperatures to compensate for the shorter heat soaking periods (Fig. 23.1). Intermittent kilns are used often in conjunction with a large tunnel kiln to give the maximum production flexibility. A constant throughput of ware is fired in the tunnel kiln while fluctuations in total factory output can be handled by operating one or more of the intermittent (periodic) kilns.

The types of pottery kiln now in use include car kilns (tunnel or intermittent), top-hat or bell types in which the ware is placed on plinths and fired by lowering on to them the 'top-hat' cover containing the refractory lining and burning units. Continuous conveyor belt kilns are also used in which the ware is set one piece high and allowed to travel at various speeds through the tunnel which is heated by oil or gas. For the firing of sanitaryware (Fig. 23.2) the British Gas Industry, collaborating with pottery kiln designers, has developed the Trent kiln on the principles of continuous conveyorisation. Ware is pushed through the short tunnels on refractory slabs. The main advantages of small cross section kilns of the Trent type include a very fast firing cycle (since there are no large masses of clayware to be heated), low temperature variations across the section of the kiln, and small capital investment for erecting a single kiln.

### Kiln furniture

When fine ceramics such as tableware and artware are being fired it is necessary to support the pieces in the kiln. Ware which is coated with glaze must not touch the shelf of the kiln car on which it is being fired other than by point contact, otherwise the melting glaze will make the product adhere to the support and it will be impossible to remove it without damage. When pottery is being fired stresses causing possible distortions develop in it, and there is a need to accommodate the stresses. In designing supports for the ware this must be taken into account (Figs. 23.3–23.4). Ceramic figures and other complicated shapes may need very elaborate supports in order to prevent deformation and sagging. Biscuit (unglazed) flatware, however, can be stacked together, using alumina as a setting material between the individual pieces as the only means of support. In the bone china and porcelain industry flatware is set individually

FIG. 23.3. High-grade refractory materials such as sillimanite and alumina must be used to make kiln furniture such as this crank assembly for supporting glazed flatware during firing. *Photo:* Courtesy J. Gimson.

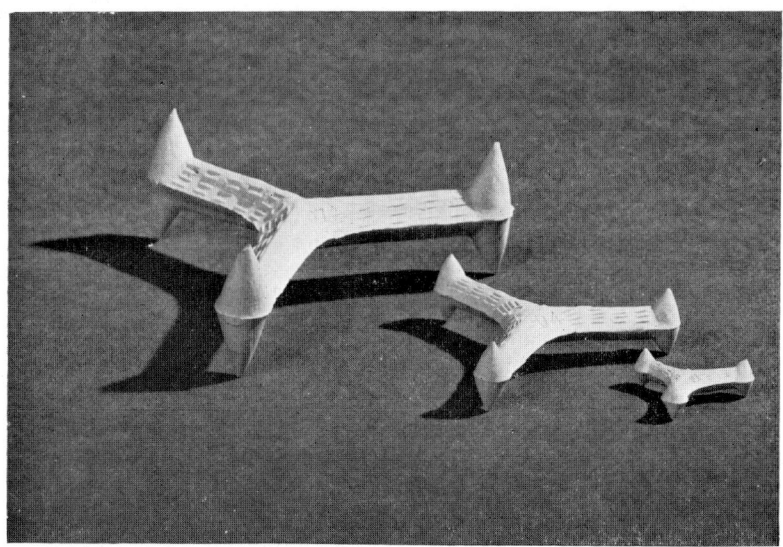

FIG. 23.4. Thousands of small high-quality refractory components such as these stilts, made by J. Gimson of Staffordshire, are used in loading a kiln truck with ware for firing. The points must be sharp to reduce the area of contact to a minimum. *Photo:* Courtesy J. Gimson.

on 'setters', that is, types of sagger, specially designed for the purpose. Porcelain and bone china which are highly vitrified during firing would otherwise sag and become distorted without the support of the setter. Alumina has largely replaced sand in the firing of ceramics as setting material because of the danger of silicosis.

Saggers are fireclay-grog boxes of various shapes in which the ware is fired (Fig. 23.1). These are not now needed so extensively because of cleaner fuels and other improved conditions. The ware is now placed in open settings on refractory slabs or bats supported by pillars or props.

The development over the last half-century in the firing of fine ceramics has been one of sagger, followed by muffle, and now open firing. Muffles are still sometimes used to prevent contact between the ware and the burner flames inside the kiln

By using pillars and slabs and various other supporting pieces various systems of shelving can be built up on the kiln car to support the ware. The setting density of refractory materials is thus quite high (Fig. 23.5).

*Materials for kiln furniture*

The refractory raw materials used in fabricating kiln furniture such as slabs, stilts, setters, etc. must be of high quality (free from iron and other fluxing agents) because the products must have a very long life. Since they are subjected to a great deal of mechanical knocking as well as thermal impact, the products must be mechanically strong and spalling resistant. Modern kiln furniture needs to have a life of several hundred cycles to be economical.

The wet moulding of fireclay-chamottes mixes to form saggers, slabs and props has now generally been replaced by the dry pressing, at very high fabrication pressures, of high-alumina materials and other sophisticated refractories. The materials now used include fused alumina with different types of bond, zircon and zirconia, silicon carbide with clay and also nitride bonds, and cordierite bodies which have a high spalling resistance.

It has long been the practice in the pottery industry to apply zircon washes to kiln furniture used in glost kilns because zircon is not readily wetted by glaze vapours. Such vapours, condensing on the kiln furniture, cause rapid failure by their slagging action. Zircon helps to prevent this and prolongs the life of the furniture. Cordierite compositions which are made by adding about 10–12 % talc to sillimanite and other high-alumina raw materials have a low thermal expansion and therefore a better resistance to thermal shock than materials not containing the talc. The cordierite phase develops during the firing of the refractories. Silicon carbide has a very high thermal-shock resistance, is very dense and mechanically very strong. However, in the firing process the silicon carbide is oxidised and may make the kiln atmosphere reducing by virtue of the presence of

FIG. 23.5.   Side baffles of refractory slabs are used in this densely set structure to protect the fine bone china from the kiln atmosphere. *Photo:* Courtesy Diamond Clay Co.

carbon monoxide. Where this is not specifically required it may cause faulty ware. Various coatings are applied to silicon carbide to prevent oxidation.

### Refractory materials firing
The linings of kilns used for firing refractories have had greater demands placed on them as a result of the need for more stable refractories by the metallurgical and other consumer industries. The quality of refractories fired in ceramic kilns depends on the maximum temperature to which they are subjected as well as the chemical and mineralogical composition. Thus, the quality of a common firebrick can be greatly improved with regard to after-contraction and slag attack by increasing its firing temperature and possibly by using chamotte that has been fired to higher temperatures.

As in the firing of all ceramics the reactions occurring during kiln firing go further to completion, hence to stability, when the temperature is raised. The demand for such new products as direct-bonded brick, mullite, and other volume-constant refractories means that tunnel kilns employed in the ceramics industry must be built of materials that can withstand higher firing temperatures. Many refractories kilns fired on oxygen-injected oil feed reach temperatures of 1 800°C; in the burner areas even higher. A typical kiln used in refractories firing uses silica (low and high duty) brick, magnesite, alumina and chrome–magnesite refractories, together with a range of insulating products (kaolin lightweight, fireclay and castable materials such as bubble-alumina concretes).

### Metal-enamel firing
Vitreous enamels, known also as metal-enamels, porcelain enamels and silicate enamels, are fired onto iron, steel, copper and aluminium at temperatures of 850–950°C. The demand on the refractories, at least in terms of heat resistance, would not, therefore, seem to be very great. However, the flame temperatures behind the muffles which are employed in the furnaces are very high, and refractories must be of a corresponding standard. The common furnaces employed in this section of the ceramics industry are muffle types (intermittent and continuous). Recently designers have been developing new types, specifically the radiant tube furnace which will tend to reduce further the refractory requirements of enamel firing processes.

Refractories employed in firing enamels include highly conducting carbides and high alumina. The ware must be protected from combustion products in order to avoid blistering and scumming faults if oil or gas fuels are used: hence the use of muffles. Muffles are made of silicon carbide, mullite and sillimanite. Heat-resistant steel has also been tried.

Vargin (1967) has described the various types of furnace used in enamel firing. In heat radiation furnaces, for example, the ware is not placed in a muffle but is placed in the firing chamber which is heated with pipes (about 8 mm wall thickness) made from heat-resistant alloys. Gas or oil is burnt inside the tubes using special burners. The advantages over conventional furnaces include faster operation (particularly with regard to cooling) and better temperature control. Electric furnaces are also employed for enamel firing. The atmosphere is clean and refractories are of the usual type employed in electric furnace design, *e.g.* aluminosilicate and silicon carbide, fused mullite and bubble alumina. The electric elements are made of molybdenum disilicide, silicon carbide and similar materials.

### Tunnel kiln car tops
The performance of the refractories on cars used in tunnel kiln firing is very important not only for the general economics of the firing technology

FIG. 23.6. Tunnel kiln car tops can be built up of firebricks (as seen here) or by using guncrete and castable refractories technology. The structure must withstand repeated firings. *Photo:* Courtesy Novosti Press Agency.

Fig. 23.7. A loaded tunnel kiln car, showing the responsibility of the refractories used in the car-top construction and the kiln furniture. *Photo:* Courtesy Novosti Press Agency.

but also to ensure the stability of the ware passing through the kiln (Figs. 23.6–23.7). Recent developments in forming the tops of the kiln cars, that is, the sections undergoing the greatest heating, have included the use of castables and guncreting methods. On top of a foundation of common firebrick it is usual to lay a layer of dense or porous refractory concrete. Shuttering is built up around the metal framework of the car, and then the refractory concrete is poured into place. The required temperature joints have to be left to accommodate expansions in heating. A typical concrete composition contains 30% calcium aluminate cement and 70% chamotte comprising about 42% <1 mm grain, 17% 1–10 mm grain, and 11% up to 30 mm grain. The concrete shrinks about 0·3%. Some operators apply vibration to the poured concrete in order to consolidate it and remove air.

Kiln car tops must be spalling resistant, have low thermal expansions, and be able to stand up to mechanical abuse as well as high temperatures under load. The precise temperatures will depend, of course, on the type of kiln in which the cars are used. However, since the products being fired

FIG. 23.8.  New kiln designs such as this sanitaryware kiln need reliable refractories because of the stresses developed when the hood is being raised and lowered. *Photo:* Courtesy Shanks Company and Carborundum.

are invariably very heavy when we take into account the car superstructure, which includes ware and kiln furniture, it is apparent that the refractoriness under load of the car top must be adequate.

Remmey (1970) has described the production of shock-resistant car blocks for making kiln car tops. He suggests that compared with solid underdecks which have been used in the past the pier subdeck, designed with shock-resistant extruded cellular blocks, makes a superior and cheap top for firing building brick.

The guncreting of tunnel kiln car bases is also claimed to produce cheaper and more resistant structures. Guncrete materials based on aluminous cement and chamotte (75 % by volume of chamotte <10 mm and 25 % cement) are being used for this purpose. The guncreting is done with a wooden shuttering fixed around the base of the car. Reinforcing wires are first fitted on the metalwork of the car in order to increase the strength of the final product. The total thickness of the refractory lining is about 250–300 mm, and there is no need for temperature joints to be included. The guncreting technique can be used for repairing corners and broken edges and also for replacing large sections of the kiln car top. For the above composition the maximum firing temperature is about 1 300°C, and it is necessary to hold the newly repaired car for three days at room temperature, periodically sprinkling with water in order to develop mechanical strength. The car is then dried for two days at 50°C before being used in the kiln. Refractories engineers, Belogrudov *et al.* (1971), who have developed this method in the Soviet Union, claim that the life of tunnel kiln cars treated in this way is five times longer than those with tops made from brick. The labour productivity of the repair job is doubled and the working conditions are improved. The costs of lining the tops of the cars are reduced because of the lower cost of the raw materials and the elimination of brickwork.

## REFERENCES AND BIBLIOGRAPHY

Belogrudov, A. G., Kalugin *et al.* (1971). Guncreting tunnel kiln car tops, *Ogneupory* No. 6, 58–61.

Dale, A. J. (1964). 'Modern Ceramic Practice'. Applied Science, London, pp. 185–92.

Fickel, F. (1970). Application of refractories in the ceramic industry, *Glas-Email-Keramo-Tech.* **21**, No. 3, 69–74.

Groom, B. and Chivers, M. A. (1971). Ware support systems, *Ceramics* **22**, Nos. 1–2.

Klose, G. R. and Kaboth, D. (1970). Refractories for vacuum furnaces, *Keram. Zeitung* No. 10, 623–8.

Lukinov, M. I. (1959). Ceramic drain pipes, *Stroiizdat*, Moscow. (Selected translations appeared in *British Clayworker*, October 1965–June 1966.)

Remmey, F. B. (1970). *Bull. Amer. Ceram. Soc.* **48**, No. 3, 266–368.

Rhodes, D. (1968). 'Kilns: Design, Construction and Operation'. Chilton Book Co., Philadelphia.

Richardson, H. M. (1969). Refractories and the effects of fuels thereon (in vitreous enamel furnaces), *Bull-Inst. Vit. Enamel* **20**, No. 3, 101–10.

Rowden, E. (1964). 'The Firing of Bricks'. Brick Development Association, London.

Vargin, V. V. (1967). 'Technology of Enamels'. Applied Science, London.

Wanie, W. (1965). Kiln furniture, *Sprechsaal* **98**, No. 20, 702–10.

*Chapter 24*

# Miscellaneous Uses of Refractories

## STEAM RAISING

Modern boilers and steam raising plant are very complex. The whole productive system in many industries often depends on their efficient operation. A large factory may use steam plant containing 200 tons of refractories. The combustion chamber in such plant may be 1 000 cubic metres in volume and temperatures may rise to 1 750–2 000°C in the burner zones. The working conditions are, therefore, very harsh. In addition to sharp variations in temperature, the materials must stand high loads, slag attack from fuel ash and other severe conditions.

The types of refractories employed depend on the types of fuel, the method of combustion, the boiler outputs and the design of the heating surfaces and the burners. Refractories technologists are particularly concerned about the type and behaviour of refractories used in the combustion chambers and the burner units. The fuels are chiefly gas, oil and pulverised coal. Refractory performance depends also on the method of slag removal (for example, wet or dry). The main types of refractories are aluminosilicate, especially high-alumina and mullite, silicon carbide, chromite, and combinations of these materials to form composites and multilayer structures. Linings working at temperatures up to 1 200°C employ firebrick and high alumina, up to 1 400°C kaolin brick, mullite, etc. while baffle coatings are sometimes made of chromite and special coatings. Other refractories employed include castables, insulation and concretes. Guncreting is a technique which is widely used in applying linings to boiler work.

Steam employed to drive steam turbines and engines can be generated by two main types of boiler: the water tube type and the fire tube type. In the former the water circulates through systems of tubes heated from outside. In the latter the heat passes inside the tubes, and is transmitted through the tube material into the water to be heated, thus producing the steam.

**Burner design**

Many different designs are used in burner construction including cyclone burners and burners for combusting liquid fuels, gas, and solids. The conditions under which refractories operate in the various types of burner also vary widely, and the particular choice of refractory and the method of applying it to the lining depend on the precise nature of the design.

**Cyclone combustion chambers**

The operating conditions of refractories in these have been discussed by Broadfield (1969) who has related the basic properties of refractory ramming compositions to their performance in the cyclone. It is found that most information in this field is chiefly of Russian and German origin.

   In cyclone burners the temperatures may rise to 1 800°C and together with this the high vortex movements of the slag and ash particles cause rapid wear of the refractories. High-alumina and silicon carbide refractories have been used here. However, according to some investigators and operators silicon carbide and aluminosilicate used individually do not guarantee the necessary service life to the combustion chambers. Chromite bodies have, therefore, been tried for the baffles of cyclone burners. The mechanism here is that the chromite coating serves only in the first period of cyclone precombustion operation, and then it is spent, the job of protecting the baffle being undertaken by a thin film of congealed slag.

   Broadfield's conclusions indicate that silicon carbide rammed linings in cyclone burners are most satisfactory, especially in view of their high thermal conductivity.

**High-pressure boilers**

These are frequently used in ships because they are very compact. In ship's boilers of all types the lining wear is very severe, the combustion temperature reaching 1 750–1 850°C for fan-blown units, and for high pressurised boilers they reach 2 000°C. High flue-gas velocities and sudden variations in temperature place great demands on the refractory linings. Most refractories employed for fan-blown boilers are unsuitable for high-pressure boilers. Double-layer linings made of an aluminosilicate base together with a lining of special silicon carbide with a silica bond have been developed. The latest information, according to Karklit *et al.* (1970) suggests that composite structures perform more satisfactorily. High-pressure boiler units in Russia are now being built with shaped refractories made of oxynitride-bonded silicon carbide, for filling the gaps formed in the lining by these shapes with a ramming body based on chamotte, 15% plastic clay, and an addition of vanadium pentoxide. The moisture content for the ramming body is about 10%. This eliminates the use of firebricks completely. Figure 24.1 shows the shaped refractories developed for this purpose.

Brickwork joints are usually the first sites to fail in boiler work and the choice of a suitable mortar is equal in importance to the choice of the bricks. In ship's boilers, for example, baffles and linings may last for less than six months at temperatures of up to 1 600°C. The failure is most generally due to weak joints which fail under vibration, irregular temperature distribution and fuel-ash attack. Cracks begin to form at the edges of the bricks denuded of mortar. Soon the whole lining caves in. One suitable mortar recommended by Misyura (1971) for ship's boilers consists of 40% water glass (modulus 2·35–2·75 and density 1·36–1·39), 32% kaolin, 16% chamotte and 12% talc (Soviet Patent No. 255823, 1969).

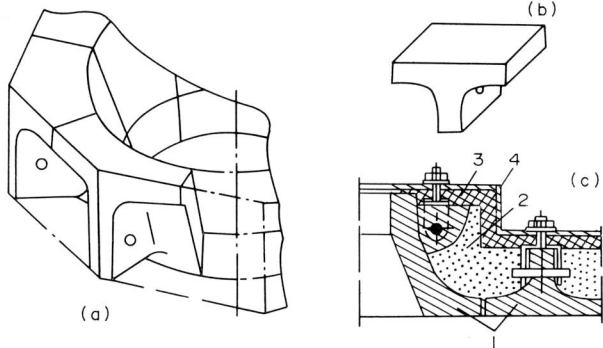

FIG. 24.1.   Lining boilers with silicon carbide; (a) blast refractories; (b) wall refractories; (c) section of lining; (1) SiC; (2) ramming body; (3) asbestos; (4) metal sheet.

## CHEMICAL INDUSTRY

The chemical engineering industry requires large quantities of chemically resistant and heat-resistant materials for constructing reactors, pipelines and other units. In addition to acid proof ceramics, such as porcelain and high-alumina compositions, the industry is now making extensive use of special refractories such as oxides, nitrides, silicides, etc. (*see* Chapters 16 and 17). Valves and pumps together with pipelines are now being constructed of materials such as boron nitride and silicon nitride. The outstanding chemical resistance and abrasion resistance of nitrides are employed in fabricating nozzles for pouring and spraying corrosive compounds. Boron carbonitride ($B_xN_yC_z$) is highly resistant to volatilisation in vacuum against many molten metals, salts and slags. It is highly electrically insulating and resistant to plasma. It is finding use in vessels containing molten borax, especially for electrolysis. Boron nitride is being used as a filler for rubber, especially microporous rubber for industrial and other purposes.

Silicon nitride is used for containing chemically active liquids and corrosion-resistant mixers. Many chemical plants now operate vessels lined with silicon nitride. Cermets, that is combinations of metals and ceramics, are also used in chemical engineering on an increasing scale. Traditional and newer types of refractories are employed by chemical engineers in the construction of high-temperature and medium-temperature heating and calcination equipment, for example in the production of lime, phosphates and other basic raw materials of the industry.

Venable (1969) has outlined the refractory requirements for ammonia plants. These materials must stand up to the adverse effects of hot hydrogen and steam, must be sound thermal insulators and also resist flames, thermal shock, and erosion. High-pressure developments have made it necessary to use low silica refractories, including high-alumina castables, bubble alumina castables and various others.

## AEROSPACE MATERIALS

Developments in recent years in space exploration and rocket technology have involved close collaboration between vehicle engineers and refractory materials technologists. The number of new materials now being used in such applications is increasing all the time. However, the nature of these materials and the discussion of their properties and specialised applications is outside the scope of the present volume. The references at the end of this chapter will give some indication of sources for detailed information on the subject. The ceramic literature and also specialist publications dealing with nuclear power technology and space research contain articles and papers on this subject. The particular types of materials use in aerospace include cermets, and other powder-technology materials, oxides, nitrides, silicides, etc. and other materials dealt with in Chapters 16 and 17. Carbon and graphite refractories are also of relevance. Fibre materials such as carbon fibres used in special components of rocket motors and jet engines, as well as in other parts of aircraft, are also of interest to the aerospace technologist.

## GAS PRODUCTION

The manufacture of town's gas from coal once demanded the production of vast quantities of refractories, especially silica brick, but new methods of producing gas (continuous reforming processes) dating from the mid 1960s and also the more recent discoveries of vast deposits of natural gas in the European area have made the development and application of refractories for gas production somewhat obsolete.

The developments and cooperation between the refractories industry and the gas industry in Great Britain were thoroughly summarised in an article for the Joint Refractories Committee of the Gas Council and the British Ceramic Research Association by Clements (1969). This report gives numerous references and emphasises the importance to the gas industry of refractories developments in the continuous retort which for many years was the chief production unit. Main problems included flaking, leakage and erosion of the units. Minor problems developed through the action of alkali on firebrick. Clements also points out that the conversion from town's gas to natural gas should not create special difficulties for refractories materials since the atmospheres inside these furnaces developed by natural gas and town's gas are identical.

The problem of silica and other refractories for coke furnaces, in which gas is also made, continues to be of importance to the refractories technologist.

## OTHER USES OF REFRACTORIES

As mentioned in previous sections of this book, refractory materials are employed by virtually every industry. Most industries involve the generation of steam, and other heat-using processes. Paper mills use refractories for sulphite regeneration; many types of incinerator design are employed in all sorts of industries, and crematoria also need to employ refractories for their social services. In the field of domestic heating, the growing use of central heating has eliminated or greatly reduced the consumption of common firebrick for firebacks and other purposes in grate construction. However, the use of domestic boilers fired on oil, solid fuel and gas requires the employment of well-made highly resistant refractory materials which will give steady if unspectacular service over many years.

## REFERENCES AND BIBLIOGRAPHY

Allbutt, M. and Dell, R. M. (1967). Chemical aspects of nitride, phosphate and sulphide fuels, *J. Nuclear Materials* **24**, No. 1, 1–20.
Broadfield, E. R. (1969). *Refractories J.* **45**, No. 8, 228–34; and No. 10, 302–11.
Clements, J. F. (1969). *Refractories J.* No. 12, 362–68.
Elston, J. (1966). Effect of radiation on ceramics, *Bull. Soc. Franc. Ceram.* No. 70, 29–39.
Gillin, L. M. (1967). Deformation characteristics of nuclear grade graphite, *J. Nuclear Materials* **23**, No. 3, 280–88.
Holden, A. N. (Editor) (1968). 'High-Temperature Nuclear Fuels'. New York.
Karklit, A. K., Krasotkina, N. I. *et al.* (1970). *Ogneupory* No. 2, 18–23.
Majourel, P. (1966). Behaviour of refractories in contact with natural gas flames, *Verres et Refractaires* **20**, No. 4, 239–42.
Misyura, K. N. (1971). *Ogneupory* No. 7, 58.
Venable, C. R. (1969). *Bull. Amer. Ceram. Soc.* **48**, No. 12, 1114–17.

# Index